Programmable Controllers

Hardware, Software, and Applications

Programmable Controllers

Hardware, Software, and Applications

George L. Batten, Jr.

Second Edition

McGraw-Hill, Inc.

New York San Francisco Washington, D.C. Auckland Bogotá
Caracas Lisbon London Madrid Mexico City Milan
Montreal New Delhi San Juan Singapore
Sydney Tokyo Toronto

Library of Congress Cataloging-in-Publication Data

Batten, George.
    Programmable controllers : hardware, software, and applications /
by George L. Batten, Jr. — 2nd ed.
        p.   cm.
    Includes index.
    ISBN 0-07-004214-4
    1. Programmable controllers. I. Title.
TJ223.P76B38 1994   93-48744
629.8'9—dc20   CIP

1 2 3 4 5 6 7 8 9 0   DOH/DOH 9 9 8 7 6 5 4

ISBN 0-07-004214-4

*The sponsoring editor for this book was Roland S. Phelps. David M.
McCandless was the managing editor and Theresa Burke was the
manuscript editor. The director of production was Katherine G. Brown.
This book was set in ITC Century Light. It was composed in Blue Ridge
Summit, Pa.*

Printed and bound by Donnelley of Harrisonburg.

*In order to receive additional information on these or any other
McGraw-Hill titles, in the United States please call 1-800-822-8158.
In other countries, contact your local McGraw-Hill representative.*

Notices

StepLadder™ **STEPS**™	Active Systems Group, Inc.
SNAP™	Automatic Timing & Controls
Blue Earth™	Blue Earth Research
Series One™ **Series Three**™ **Series Five**™ **Series Six**™ **Series 90**™ **Workmaster**™	GE Fanuc Automation
S9000™	Honeywell, Inc.
ServoPro™	Industrial Indexing Systems
IBM®	International Business Machine Corporation
SY/MAX™ **SY/NET**™	Square D Company
Little Giant™ **Little PLC**™ **SmartBlock**™ **Dynamic C**™	Z-World Engineering

For Jeanne, Brigid, Katie, Jason, and Reilly

Contents

Appendices

Acknowledgments

Writing this book was easier because of the technical and product information provided by the people listed here. I am grateful for their assistance.

Jeanne Batten; John Hall, I&CS Magazine; Kirby McManus, Active Systems Group, Inc.; Carolyn Pennington, Adatek; Peter Matushin, Aromat Corporation; ASC Industries, Inc.; Sue Linkenhoker, Automatic Timing & Controls Company, Inc.; Thomas Hiniker, Blue Earth Research; A. P. Burns, Cegelec Industrial Controls; Lynne Nadolsky, Divelbiss Corporation; Lynna Carnett, Eagle Signal Controls; Jack Luft, Entertron Industries, Inc.; Mike McEntee, Festo Corporation; Ronald Williams, GE Fanuc Automation; Fritz Palas, Honeywell, Inc.; Michael Sullivan, ICON Corporation; Idec Corporation; Edward Steiner, Industrial Indexing Systems; Robin Chang, International Parallel Machines, Inc.; Rick Carlson, Klockner-Moeller; Minarik Electric Company; Frank Abbruscato, Modicon, Inc.; Ian Clark, Precision MicroControl Corporation; Tom Antonino, Square D Company; Lynn Alderson, Systems Engineering Associates, Inc.; Ronald Dawson, Toshiba International Corporation; Carole Fenzl, Triconex Corporation; A. J. Byam, UTICOR Technology, Inc.; Douglas Dillie, Westinghouse Electric Corporation; and Kelli Binsky, Z-World Engineering.

Introduction

When this book was first published in 1988, it was the only book on programmable controllers (PCs)[*] written for an audience who had no previous knowledge of programmable controllers. Its continued success confirms the need for such a book.

In 1968, the state of the art in control technology was the relay control system. Shortly thereafter, the programmable controller became the state of the art in control technology. Since then, programmable controllers have made rapid inroads into virtually every manufacturing business and many service businesses. The relatively rapid integration of the programmable controller into the manufacturing sector has been called "the second industrial revolution." And the revolution is far from over.

It is obvious from the foregoing that the successful technician, engineering aid, engineer, or supervisor involved in nearly any manufacturing business needs a basic knowledge of the way in which programmable controllers are used. In response to this demand for knowledge about programmable controllers, several good books have appeared. These books tend to go into great detail concerning programmable controller hardware, criteria for selection, and maintenance. This is information that the working engineer needs to know. Unfortunately, the first-time reader of these books is lost because each of these books assumes previous knowledge of programmable controllers on the part of the reader. Thus, technicians, engineering aids, engineers, and supervisors who need an introduction to programmable controllers are confronted with a dilemma: the acquisition of basic knowledge about programmable controllers depends upon the existence of a basic knowledge about programmable controllers. Where does the uninitiated reader turn for this basic knowledge?

Thus, the purpose of this second edition remains the same as that of the first edition: this book provides the reader with that basic understanding of programmable controllers.

The purpose of the book then defines the audience of the book. It is meant for technical people (technicians, engineering aids, and engineers) and supervisory personnel who require a basic understanding of programmable controller operations

---

[*] Although the acronym PC often means personal computer, in this book, PC stands for programmable controller, unless otherwise noted.

and uses. No special knowledge, skill, or experience is necessary. A basic under-standing of electricity/electronics is, however, assumed. If you understand how a re-lay works, your level of electricity/electronics education is sufficient. And if you have a passing familiarity with personal computers, this book will be a breeze.

Since 1988, programmable controllers have become more pervasive in the process industries. At the same time, they have become smaller, have increased in memory size, have become more functional (especially in terms of compatibility with intelli-gent I/O modules), and have become more user-friendly. The manufacturers of pro-grammable controllers have also taken more care in preparing their user manuals. I have read many of these manuals, and I believe the manufacturers now assume less knowledge on the part of the users of these manuals than they did in 1988. Thus, the uninitiated can learn more, with less confusion, from suppliers' literature than in the past.

Still, if you intend to learn the basics of programmable controllers from the sup-pliers' literature alone, you face a formidable task. A glance at appendix A, which is a listing of domestic PC suppliers, hints at the challenge of this approach. A large number of suppliers exist, and their literature varies in terms of user-friendliness.

Hence, the second edition of this book. As with the first edition, the specific ob-jectives remain. A reader previously unfamiliar with programmable controllers should be able to do the following upon completion of this book:

- Understand the basic components of programmable controller systems

- Reasonably understand the languages used to program these devices

- Understand how programmable controllers are used in closed-loop feedback con-trol schemes (especially proportional-integral-derivative control)

- Understand how programmable controllers communicate with peripheral devices and other programmable controllers in networks

- Read a technical data sheet on a programmable controller and understand the ca-pabilities and limitations contained therein, without being overwhelmed by the sales pitch

- Pick up any of the advanced texts listed in the bibliography and benefit from read-ing them.

In other words, this book will not turn the previously uninitiated reader into an ex-pert on programmable controllers, but it will give the reader a solid foundation on which to build expertise.

The book divides naturally into sections. The first section includes chapters 1 through 5 and provides background material. Chapter 1 defines the programmable controller and compares programmable controllers with personal computers. The history of the programmable controller is recounted, and an incomplete list of in-dustries employing programmable controllers is given. Chapter 2 introduces logic circuits. These are the circuits that form the building blocks of the programmable controller's central processing unit. The logic operations introduced in this chapter also appear later in the form of a programming language. Chapter 3 introduces Boolean algebra, the natural mathematics of the circuits introduced in chapter 2.

Chapter 4 deals yet again with numbers. This time, the focus is number systems. Our own decimal number system is compared with systems that find widespread use in programmable controllers and personal computers. These are the binary, octal, and hexadecimal number systems. Because the programmable controller's microprocessor actually uses the binary system, that system is highlighted. Binary arithmetic is developed, and circuits for the performance of binary arithmetic are described. Finally, binary number codes are introduced. Chapter 5 examines the relay and how it was used in control applications. Because the programmable controller was developed to replace relay control systems, this chapter is essential to understanding programmable controllers. The ubiquitous relay-ladder diagram makes its first appearance in this chapter.

Chapters 6, 7, and 8 form the second section of the book. This section addresses programmable controller hardware. Chapter 6 covers the central processing unit: the power supply, microprocessor, and memories. Chapter 7 describes input/output interfaces, the devices that allow the programmable controller to adapt to the outside world. This chapter also contains a thorough discussion of proportional, proportional-integral, and proportional-integral-derivative control. Finally, chapter 8 covers peripherals, or external devices (such as printers), that are connected to the programmable controller.

Chapter 9 forms the third section of the book. If you own a personal computer, you know that it is of very little use without appropriate software, or programs. The business of programming a programmable controller is complicated by two factors: a variety of programming languages exist, and the most popular ones are based on the logic used to wire relay control systems. Relay-ladder diagrams are a holdover from the days before programmable controllers and tend to confuse younger engineers who never used relays in control. High-level computer languages tend to confuse the older engineers, who still remember the days when relays controlled plant processes. Both low-level (relay-type) and high-level (computer-type) programming languages are described in this chapter. A comparison of both types of languages points out the similarities (and disparities) between the two.

The fourth section of the book is chapter 10, and the fifth section is chapter 11. Chapter 10 provides practical examples of how programmable controllers are used in process measurement and control, and chapter 11 describes how they communicate with each other in local area networks. These chapters round out the basic introduction to programmable controllers.

The sixth section includes chapter 12, the glossary, the bibliography, and the appendices. This is a section of reference material. Chapter 12 provides an overview of the programmable controllers available in the United States at the time of this second edition. In light of the information contained in the first 11 chapters of the book, it is interesting to note the diversity of programmable controllers and systems. The glossary is convenient for quick reference. The bibliography suggests texts for further reading, as well as courses and periodicals.

The appendices provide a variety of material:

- Names and addresses of suppliers of programmable controllers
- Introductory information on a new, high-level programming language

- The relay STEPS™ relay ladder programming language
- A summary of Boolean algebra
- The ASCII code
- A summary of relay-ladder and Boolean programming symbols
- Selected literature on commercial programmable controllers.

The suppliers' literature included in appendix G has been selected to illustrate not only the full-sized programmable controllers used for large control schemes, but also the low-end (small, inexpensive) programmable controllers that are causing quite a stir these days.

Programmable controllers have constituted the second industrial revolution. The successful technician, engineering aid, engineer, or supervisor involved in nearly any manufacturing business needs a basic knowledge of the way in which programmable controllers are used. It is my hope that this edition will satisfy that need, much as the first edition did five years ago.

Now you know why the book was written, who should read it, and what to expect. It is time now to forge ahead and to begin learning about these remarkable, indispensable devices. Good luck!

George L. Batten, Jr.

Background

What is a programmable controller? Briefly, it is a digital electronic device that meets the following three criteria:

- It has a programmable memory, in which instructions can be stored.
- The instructions stored in the memory are used to implement various functions, such as logic, sequencing, timing, counting, and arithmetical functions.
- The various functions are used to control machines or processes.

The control functions of the programmable controller (PC) are accomplished through input/output modules, which can be either analog or digital.

From this description, it would seem that a personal computer can be used as a programmable controller. Indeed, personal computers can be used as programmable controllers, as shown in Figure 1.1, in which a personal computer is used as a home-security programmable controller. In this example, three types of sensing elements are connected to the personal computer via an input interface module:

- A photocell, which measures light level.
- A smoke detector (or several smoke detectors).
- Various security switches on doors and windows.

The memory of the personal computer has been programmed with instructions to be implemented by the microprocessor. For example, if the light level in a room falls below a predetermined value, the microprocessor is instructed to turn on the room lights. If a signal is received from the smoke detector, the microprocessor is instructed to sound an alarm, dial the fire department, and deliver a prerecorded message. By performing these specified functions, the personal computer controls light switches, burglar and fire alarms, and telephone communications via the output in-

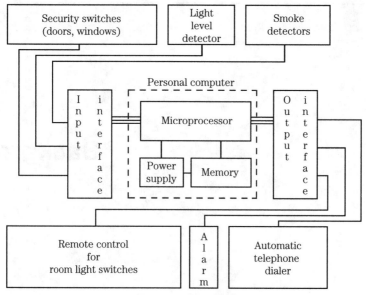

Figure 1.1 Personal computer used as a programmable controller.

terface module. Thus, the personal computer in this application satisfies the three previously listed criteria and functions as a programmable controller.

Although personal computers can function as programmable controllers, they are generally not used for this application. There are several reasons for this: personal computers are extremely flexible devices that can be used to perform calculations, plot graphs, play games, and instruct children. Why should someone tie up such a flexible and useful device for the mundane purpose of home security? Why not develop a microprocessor-based home-security programmable controller designed to perform this series of tasks only?

Personal computers are also not used as programmable controllers because of the environment in which the control functions must often be performed. Many manufacturing plants do not have the carefully controlled temperatures and humidities that computers require. Additionally, plants that use pumps and motors generally experience electrical interference problems. Personal computers are not designed to perform in such rugged environments, but programmable controllers are.

In summary, the programmable controller does share features with the personal computer: it is a digital, electronic, programmable device that controls machines or processes. The programmable controller differs from the personal computer, however, in that the controller is more rugged and somewhat less flexible than the personal computer.

The History of the Programmable Controller

The history of the programmable controller is not extensive. In fact, it goes back only as far as 1968, to the Hydramatic Division of General Motors Corporation. The mass manufacture of automobiles (and automobile components) involves many machines, all of which must be controlled. Consider, for example, the boring machine. For the boring machine to bore a hole in the center of a piece of metal, a control mechanism is necessary to prevent the boring machine from operating until the piece of metal is properly aligned. This control mechanism must also actuate the boring machine when the metal is aligned and then reverse the direction of the bit when boring has been completed. Prior to 1968, this control function was performed by control relays. (How relays are used as control devices are described in chapter 5.)

Control relays were effective, but they suffered from several disadvantages. First, relays are capable only of on/off control, so many relays are needed to design complicated control systems, making the relay control scheme quite expensive. Control relays can also be bulky, so a control system requiring many relays takes up much floor and cabinet space. They are power-hungry too, and this high power consumption results in heat generation. When a relay fails, either through an opening of the coil or a pitting of the contacts, troubleshooting or locating the failed relay is difficult. Worst of all, relays are hardwired. Any change in the control program requires relays to be rewired. Rewiring is extremely costly, both in terms of labor and plant downtime, and is of critical importance, because (in the case of General Motors) relay control systems must be changed each year for differences in the current year's model.

Thus, General Motors had sufficient incentive to eliminate the relay-control system. In 1968, GM specified design criteria for a programmable-logic controller (PLC) to replace the relay-control system. The specified criteria required that the PLC be:

- Easily programmed or reprogrammed with a minimum of downtime (or loss of service).

- Easily maintained (i.e., modular and self-diagnosing).

- Rugged enough to operate in an industrial environment.

- Able to consume less power and require less cabinet and floor space than the relay control system.

- Competitive in cost.

Additional specifications required that the PLC have expandable memory, communicate with data collection systems, and accept 120-volt ac signals.

The specification attracted the attention of several manufacturers of control equipment, and the results of their efforts were the first-generation programmable controllers. Those first programmable controllers were quite primitive when evaluated by today's standards. They, like the relays they replaced, simply functioned as on/off controllers. The reduced space requirements and power-consumption specifications were met, however, and the PCs did have primitive self-diagnostic indicators that aided during troubleshooting. These features caused these primitive PCs to be widely used, and with widespread use came improvements. Developments in microprocessor technology also translated into more flexible and powerful PCs.

As time progressed, PCs acquired the capacity to do arithmetic, manipulate data, and communicate more efficiently with the programmer. Following innovations in microprocessor and memory technology, PCs acquired larger memory capacity and the ability to communicate with other controllers or a master *host computer*. With the development of analog control, PCs were able to move past simple on/off control to more complex schemes, including proportional-integral-derivative (PID) control. With these developments, the cost of PCs steadily decreased.

In summary, today's PCs are economical, user-friendly devices that have benefited from advances in microprocessor and memory technologies. Today's PCs are capable of performing complicated control routines and can communicate with other PCs and host computers in sophisticated control networks.

PC Applications

Very few industries exist today that do not employ programmable controllers. Almost every business in the manufacturing sector, and many in the service sector, use PCs in abundance. A very incomplete list of industries using PCs include the following: aerospace, automotive, bottling and canning, chemicals, entertainment, food and beverage, gas and petroleum, lumber, machining, metals, mining, packaging, petrochemicals, plastics, power, pulp and paper, rubber, and transportation. In these industries, PCs perform a variety of control functions, from weighing, conveying, and handling materials to drilling, boring, converting, and packaging.

It is little wonder that the relatively rapid integration of the PC into the manufacturing sector has been called the second industrial revolution. And the revolution is far from over.

2

Introduction to Logic Circuits

Chapter 1 presented three criteria that programmable controllers must meet: the presence of a programmable memory for the storage of instructions, the use of the instructions to implement various functions, and the use of these functions to control machines or processes. To meet these criteria, logic circuits are used. In this chapter, the elements of circuit logic are outlined, and the basic logic circuits (AND, OR, and NOT gates) are discussed.

The AND Gate

If you rent a safety deposit box at the local bank, you are familiar with the concept of the AND gate. To enter your safety deposit box, two keys are required: the renter's key and the banker's key. If both keys are inserted and turned at the same time, the safety deposit box is opened.

The electrical equivalent of two keys is the AND function; it is shown in Figure 2.1. In this simple model, the function to be performed is to turn on the lamp. The lamp turns on when two conditions are satisfied simultaneously: switch 1 is closed, and switch 2 is closed.

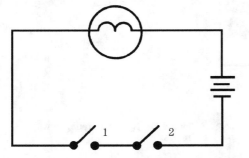

Figure 2.1 Simple model of an AND gate.

A variation on the circuit in Figure 2.1 can be created using relays, as shown in Figure 2.2. In this figure, the light is turned on when both relays are energized simultaneously, i.e., when a signal or voltage is applied to both relays. Thus, the AND function is performed only when a voltage is applied to relay 1 *and* relay 2.

The circuit in Figure 2.2 is redrawn in Figure 2.3. In this figure, the relays and battery have been enclosed by a dashed line, which is the symbol for an AND gate. Modern semiconductor AND gates do not comprise relays, of course, but the principles of operation are the same as for the relay circuit shown in Figure 2.3: when voltages are applied to all inputs of the AND gate, an output voltage is obtained.

The modern semiconductor AND gate is a *binary* device—it recognizes only two states of existence: off or on. A more accurate statement, however, would be that the device recognizes either low or high voltages because digital logic circuits generally use +5 volts dc as the on level and 0 volts dc as the off level. The off state (also known as the low or false state) is symbolized by the number 0, and the on state (also known as the high or true state) is symbolized by the number 1. Thus, if any input to the AND gate is 0, the output is 0; if all inputs are 1, the output is 1. Remember that 0 and 1 represent off and on, respectively, and the number 1 usually corresponds to a voltage of +5 volts dc.

The AND gate is not limited to two inputs, and three or more inputs to an AND gate are not uncommon. Two-input AND gates can be used in combination to build even larger input AND gates, such as, four-, six-, or eight-input AND gates, as shown in Figure 2.4.

A complete description of the outputs of an AND gate for all possible sets of inputs can be represented conveniently using a *truth table*. A three-input AND gate truth table is shown in Figure 2.5. Each horizontal line of the truth table lists a possible combination of inputs, along with the output for that combination of inputs. For example, the first line of the truth table shows that, for three false inputs, the output is false. Scanning down the table, one sees that the output is true only when *all* inputs to the AND gate are true.

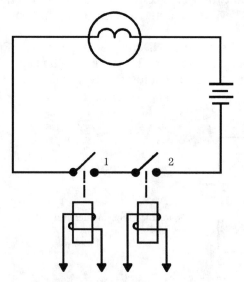

Figure 2.2 AND gate redrawn with relays.

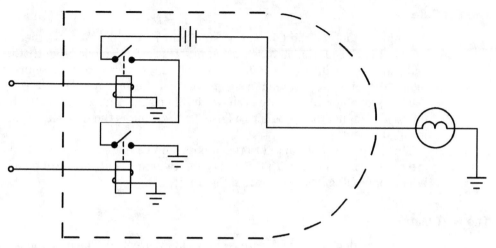

Figure 2.3 Redraw of the AND gate with relays.

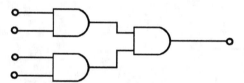

Figure 2.4 A four-input AND gate constructed from three two-input AND gates.

A
B   Y
C

AND truth table

Inputs			Output
A	**B**	**C**	**Y**
0	0	0	0
0	0	1	0
0	1	0	0
1	0	0	0
1	1	0	0
0	1	1	0
1	0	1	0
1	1	1	1

Figure 2.5 Truth table of a three-input AND gate.

The OR Gate

A simple model for the OR gate is shown in Figure 2.6. When either switch 1 or switch 2 is closed, the lamp is turned on. This simple model, like that of the AND gate, can be modified to include the use of relays, as is shown in Figure 2.7. In this figure, the lamp is activated if a signal is supplied *either* to relay 1 or relay 2.

The symbol for an OR gate is shown in Figure 2.8, along with its truth table. The only set of inputs that produces a false, or 0, output is the first set, the set comprising all 0 inputs.

The OR gate is not limited to two inputs, and OR gates with three or more inputs are not uncommon. As with the AND gate, two-input OR gates can be used to build larger input OR gates, as shown in Figure 2.9.

The NOT Gate

A simple model for the NOT gate is shown in Figure 2.10. The function performed by the circuit is to supply an ac voltage to the lamp. The relay used in the circuit is normally closed. When an on signal is supplied to the relay, the contact opens, and the lamp switches to the off state. The output of the circuit is the opposite, or inverse, of the input to the relay.

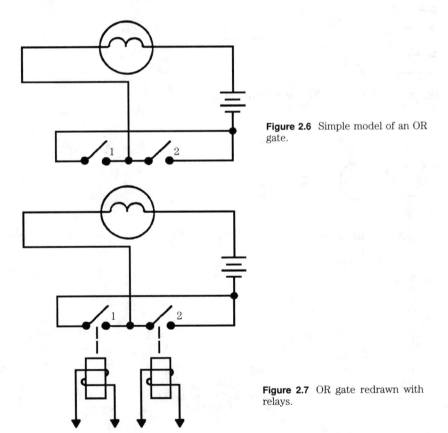

Figure 2.6 Simple model of an OR gate.

Figure 2.7 OR gate redrawn with relays.

OR truth table

Inputs		Output
A	**B**	**Y**
0	0	0
0	1	1
1	0	1
1	1	1

Figure 2.8 Truth table of a two-input OR gate.

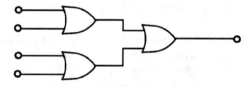

Figure 2.9 A four-input OR gate constructed from three two-input OR gates.

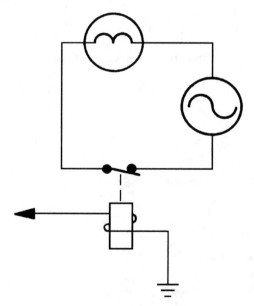

Figure 2.10 Simple model of a NOT gate.

The symbol for the NOT gate is shown in Figure 2.11, along with its truth table. The truth table is simple: the output is always the inverse of the input. If the input is 1, the output is NOT 1, or 0, and vice versa. This characteristic is why the NOT gate is often called an *inverter*.

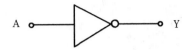

NOT truth table

Inputs	Output
A	**Y**
0	1
1	0

Figure 2.11 Truth table of a NOT gate.

Other Logic Gates

Three other logic gates need to be mentioned. These gates, the NAND, NOR, and EXCLUSIVE-OR (XOR) gates, are *composite* gates; that is, they comprise combinations of AND, OR, and NOT gates.

The principle of operation of the NAND gate, as well as its symbol and truth table, is shown in Figure 2.12. The NAND gate is essentially an AND gate followed by an inverter—NOT AND, or NAND. To obtain the NAND gate truth table, simply invert the output column of the AND gate truth table. In other words, the output of the NAND gate is 1 if *any* input is 0. If all inputs are 1, the output is 0.

NAND gate equivalent circuit

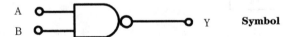

Symbol

Truth table

Inputs		Output
A	**B**	**Y**
0	0	1
0	1	1
1	0	1
1	1	0

Figure 2.12 The NAND gate.

The NOR gate is similar in construction. It is essentially an OR gate followed by a NOT gate. Inverting the output column of the OR gate truth table yields the NOR gate truth table, as shown in Figure 2.13. The output of the NOR gate is 1 only when *all* inputs to the NOR gate are 0.

The EXCLUSIVE-OR (XOR) gate is more complicated than the NAND and NOR gates. It is the equivalent of one OR, two AND, and two NOT gates, connected as shown in Figure 2.14. The only difference between the OR and XOR gates can be seen in the last line of the truth table in Figure 2.14. While the OR gate gives an output of 1 when all inputs are 1, the XOR gate gives an output of 0 when all inputs are 1.

Positive and Negative Logic

In all the preceding examples, the 1 state has been represented with a more positive voltage than the 0 state. For example, as stated previously, the 1 state usually corresponds to a voltage of +5 volts dc, while the 0 state corresponds to 0 volts dc. This type of logic, in which the 1 state is represented by a more positive voltage than the 0 state, is called *positive logic*.

A second type of binary logic called negative logic also exists. With negative logic, the 1 state is represented by the less positive voltage. For example, in a negative logic device, the 1 state might correspond to 0 volts dc, while the 0 state would correspond to +5 volts dc. Usually, positive logic is the prevalent logic; however, sometimes it is more convenient to work with negative logic. You should be aware of its existence.

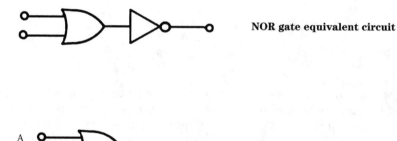

NOR gate equivalent circuit

Symbol

Truth table

Inputs		Output
A	B	Y
0	0	1
0	1	0
1	0	0
1	1	0

Figure 2.13 The NOR gate.

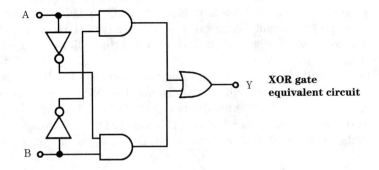

XOR gate
equivalent circuit

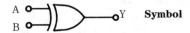

Y   **Symbol**

Truth table

Inputs		Output
A	**B**	**Y**
0	0	0
0	1	1
1	0	1
1	1	0

Figure 2.14 The EXCLUSIVE-OR (XOR) gate.

Summary

The logic gates described in this chapter perform basic logic functions and are the building blocks of programmable controllers. In later chapters, these gates are used to perform control functions that range from the simple to the complex.

3

Boolean Algebra

Mathematicians are notorious for inventing new forms of mathematics that have no immediate practical applications but later prove very useful. George Boole (1815–1864), the English mathematician and logician, was no exception. He developed an algebra of sets that bears his name, *Boolean algebra*. The symbolism of set theory is confusing to most nonmathematicians, so a complete definition of Boolean algebra and its postulates is not attempted here. It is sufficient to note that Boolean algebra provides a convenient shorthand for describing logic operations, particularly those operations of the logic gates discussed in chapter 2. This chapter introduces the elementary principles of Boolean algebra.

Boolean Expressions

In chapter 2, the letters A, B, C, and so on symbolized inputs to the logic gates, while the letter Y designated the output of the logic gates. That convention is maintained in this chapter.

Three basic Boolean expressions are of interest to those who work with logic circuits: the multiplication sign, the addition sign, and the horizontal bar above a letter or combination of letters.

The multiplication sign ($\times$, $\bullet$, or letters written together with nothing between them) is symbolic shorthand for the AND operation. Thus, the equation describing the three-input AND gate shown in Figure 3.1 is the following (shown all three ways):

$$Y = A \times B \times C \tag{3-1a}$$

or

$$Y = A \bullet B \bullet C \tag{3-1b}$$

or

$$Y = ABC \qquad (3\text{-}1c)$$

This mathematical definition can be seen clearly by examining the various horizontal rows in the truth table. In the first row, $A = B = C = 0$. When multiplied together, the result is $Y = 0$. The same is true for the second row, $0 \times 0 \times 1 = 0$. This solution holds throughout the truth table. When any input is zero, the output is zero. The last row of the truth table provides the only nonzero output. When $A = B = C = 1$, their product is $Y = 1$.

The addition sign is symbolic shorthand for the OR operation. Thus, the equation describing the two-input OR gate shown in Figure 3.2 is the following:

$$Y = A + B \qquad (3\text{-}2)$$

AND truth table

Inputs			Output
A	**B**	**C**	**Y**
0	0	0	0
0	0	1	0
0	1	0	0
1	0	0	0
1	1	0	0
0	1	1	0
1	0	1	0
1	1	1	1

Figure 3.1 Truth table of a three-input AND gate.

OR truth table

Inputs		Output
A	**B**	**Y**
0	0	0
0	1	1
1	0	1
1	1	1

Figure 3.2 Truth table of a two-input OR gate.

Again, this can be seen by examining the rows in the truth table. If $A = B = 0$, then $A + B = 0$. For any nonzero input, $A + B = 1$. And, when both inputs are 1, the output is 1. In short, if the sum of the inputs is 1 (or more than 1), the output is 1.

The horizontal bar above a letter is symbolic shorthand for the NOT operation. Thus, the equation describing the inverter shown in Figure 3.3 is:

$$Y = \overline{A} \tag{3-3}$$

This definition is shown in the truth table. Y is always NOT A, or $\overline{A}$.

Boolean Algebra

In the previous chapter, you saw how AND, OR, and NOT gates could be combined to form NAND, NOR, and XOR gates. The Boolean expressions presented in the previous section can also be combined to describe these composite gates. The rules for combining Boolean expressions are given in this section.

As with "ordinary" algebra, Boolean algebra obeys the two following commutative laws:

$$A + B = B + A \tag{3-4}$$

$$AB = BA \tag{3-5}$$

In other words, the order of two inputs to be added (ORed) or multiplied (ANDed) is irrelevant. Boolean algebra also obeys the following two associative laws:

$$A + (B + C) = (A + B) + C \tag{3-6}$$

$$A(BC) = (AB)C \tag{3-7}$$

This law simply means that, with more than two inputs, the grouping of the inputs to either an OR or an AND gate is irrelevant.

The commutative and associative laws given above describe "pure" logic operations; i.e., purely AND operations or purely OR operations. The mixing of these operations is covered by the distributive law:

$$A(B + C) = AB + AC \tag{3-8}$$

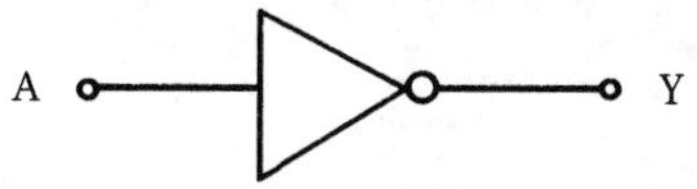

A o—————[>o———o Y

NOT truth table

Input	Output
A	**Y**
0	1
1	0

Figure 3.3 Truth table of a NOT gate.

While this law is the same as standard algebra, it is worthwhile to read equation 3-8, substituting the words *and* and *or* for the multiplication and addition signs, respectively: "A and (B or C) equals A and B, or A and C." This approach emphasizes that Boolean algebra is simply a mathematical way of stating a logical sentence.

Another distributive law is best demonstrated with logic gates. Consider the following equation:

$$Y = A + BC \qquad\qquad (3\text{-}9)$$

The logic circuit and truth table for this equation are shown in Figure 3.4. From the truth table, note that $Y = 1$ when $A = 1$, or when both B and $C = 1$. Now consider the following equation:

$$Y = (A + B)(A + C) \qquad\qquad (3\text{-}10)$$

The logic circuit and truth table for this equation are shown in Figure 3.5. The truth tables for circuit 3.4 and 3.5 are identical. Because the truth tables are identical, and equations 3-9 and 3-10 are simply Boolean equations that can be used to generate the truth table, the two equations must therefore be equal. In other words, we have another distributive law:

$$A + BC = (A + B)(A + C) \qquad\qquad (3\text{-}11)$$

Note that this definition is not the result you would obtain with standard algebra. Using the traditional rules of algebra, expanding the right side of equation 3-11 results in the following:

$$(A + B)(A + C) = AA + AC + BA + BC \qquad\qquad (3\text{-}12)$$

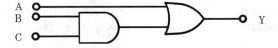

A
B
C

Y

Truth table

Inputs			Output
A	**B**	**C**	**Y**
0	0	0	0
0	0	1	0
0	1	0	0
1	0	0	1
0	1	1	1
1	1	0	1
1	0	1	1
1	1	1	1

Figure 3.4 Logic circuit and truth table for $Y = A + BC$.

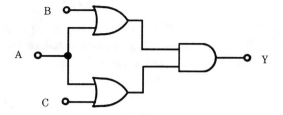

Truth table

Inputs			Output
A	**B**	**C**	**Y**
0	0	0	0
0	0	1	0
0	1	0	0
1	0	0	1
0	1	1	1
1	1	0	1
1	0	1	1
1	1	1	1

Figure 3.5 Logic circuit and truth table for $Y = (A+B)(A+C)$.

This expanded equation, however, yields the same truth table found in Figures 3.4 and 3.5. Obviously, if $A = 1$, then $AA = 1$, and if $A = 0$, then $AA = 0$. Thus, $AA = A$ in Boolean algebra. Using this identity and the commutative laws (equations 3-4 and 3-5), equation 3-12 can be rewritten as the following:

$$(A + B)(A + C) = A + AB + AC + BC \qquad (3\text{-}13)$$

Using the distributive law, equation 3-8, the first three terms on the right side of the equation can be rewritten as one term, $A(1 + B + C)$:

$$(A + B)(A + C) = A(1 + B + C) + BC \qquad (3\text{-}14)$$

Consider the term $A(1 + B + C)$. The values of B or C do not matter because a 1 is also in the parentheses. The value of the term $A(1 + B + C)$ is always determined by the value of A: If $A = 1$, the entire term must have the value of 1. If $A = 0$, the term must have the value of 0. We can thus replace the term $A(1 + B + C)$ with A, and equation 3-14 becomes the following:

$$(A + B)(A + C) = A + BC \qquad (3\text{-}15)$$

which is equivalent to equation 3-11.

In the preceding derivation of equation 3-15, two identities were discovered: any input multiplied (ANDed) by itself is simply the input, and any input multiplied (ANDed) by a term of added (ORed) inputs that includes the number 1 is simply the original input. In other words,

$$AA = A \tag{3-16}$$

$$X(1 + A + B + C + \ldots) = X \tag{3-17}$$

These two identities can be combined to yield the law of absorption:

$$A(A + B) = A \tag{3-18}$$

If we expand the left side of equation 3-18, we have the following:

$$A(A + B) = AA + AB \tag{3-19}$$

Applying equation 3-16 yields the following:

$$A(A + B) = A + AB = A(1 + B) \tag{3-20}$$

Applying equation 3-17 to equation 3-20 results in equation 3-18.

The laws of Boolean algebra governing inversion deserve special mention. Figure 3.6 gives the logic circuit and truth table for $Y = \overline{(A + B)}$. Note that the operation in parentheses, A OR B, is performed first, followed by the inversion. The intermediate output, Y', gives the output one would expect from the OR gate; however, the final output, Y, is the inverted (NOTed) value of Y'. Now consider Figure 3.7, which gives the logic circuit and truth table for $Y = \overline{A}\,\overline{B}$. Note that the NOT operation is performed prior to performing the AND operation. The two truth tables are the same, which means that the two Boolean equations are identical:

$$\overline{(A + B)} = \overline{A}\,\overline{B} \tag{3-21}$$

This identity is known as one of *DeMorgan's Laws.*

Another of DeMorgan's laws can be derived fairly easily. Figure 3.8 shows the logic circuits and truth table for $Y = \overline{(AB)}$ and $Y = \overline{A} + \overline{B}$. Again, both circuits follow the same truth table, so they must be equal, yielding another of DeMorgan's laws:

$$\overline{(AB)} = \overline{A} + \overline{B} \tag{3-22}$$

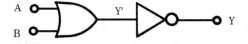

Truth table

Inputs		Outputs	
A	**B**	**Y'**	**Y**
0	0	0	1
0	1	1	0
1	0	1	0
1	1	1	0

Figure 3.6 Logic circuit and truth table for $Y = \overline{(A+B)}$.

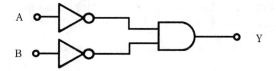

Truth table

Inputs		Output
A	**B**	**Y**
0	0	1
0	1	0
1	0	0
1	1	0

Figure 3.7 Logic circuit and truth table for $Y = \bar{A}\,\bar{B}$.

$$Y = (\overline{AB})$$

$$Y = \overline{A} + \overline{B}$$

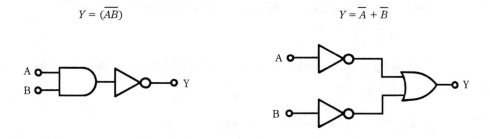

Truth table

Inputs		Output
A	**B**	**Y**
0	0	1
0	1	1
1	0	1
1	1	0

Figure 3.8 Demonstration of DeMorgan's law.

The order in which Boolean operations are performed is of extreme importance. Generally, operations in parentheses are performed first. Thus, on the left side of equation 3-22, the AND operation is performed prior to the NOT operation. When no parentheses are present, the priority order for operations are NOT, AND, and then OR. Thus, on the right side of equation 3-22, the NOT operation is performed on both inputs prior to the OR operation.

When using inversion, be careful to note the extent of the horizontal line above the symbols. As an exercise, you should try to demonstrate that $(\overline{AB})$ is not equal to $\overline{A}\,\overline{B}$.

In the preceding derivations of DeMorgan's laws (equations 3-21 and 3-22) only two inputs were considered. DeMorgan's laws, however, hold for more than two inputs:

$$\overline{(A + B + C + D + E)} = \overline{A}\,\overline{B}\,\overline{C}\,\overline{D}\,\overline{E} \tag{3-23}$$

$$\overline{(ABCDE)} = \overline{A} + \overline{B} + \overline{C} + \overline{D} + \overline{E} \tag{3-24}$$

When an input is inverted twice, its value is that prior to inversion:

$$\overline{\overline{A}} = A \tag{3-25}$$

where the double bar represents inversion twice. Obviously, this holds true for AND and OR operations on inputs:

$$\overline{\overline{(AB)}} = AB \tag{3-26}$$

$$\overline{\overline{(A + B)}} = A + B \tag{3-27}$$

DeMorgan's laws can also be applied to twice-inverted inputs:

$$\overline{\overline{(A + B + C)}} = \overline{\overline{A}}\,\overline{\overline{B}}\,\overline{\overline{C}} = A\,B\,C \tag{3-28}$$

This section concludes with three useful identities. Consider the following:

$$Y = AB + A\overline{B} \tag{3-29}$$

Using the distributive law, it can be written as

$$Y = A(B + \overline{B}) \tag{3-30}$$

If $B = 0$, then $\overline{B} = 1$, and vice versa, so the value of Y is determined by the value of A. In other words,

$$AB + A\overline{B} = A \tag{3-31}$$

The second identity is derived as follows:

$$Y = A + \overline{A}B \tag{3-32}$$

If $A = 1$, then $Y = 1$. If $A = 0$, then $\overline{A} = 1$, and the value of Y is determined by the value of B. In other words, Y equals either A or B. Thus,

$$A + \overline{A}B = A + B \tag{3-33}$$

For the third identity, consider the following:

$$Y = AB + AC + B\overline{C} \tag{3-34}$$

The truth table for this Boolean equation is as follows:

A	B	C	Y
0	0	0	0
0	0	1	0
0	1	0	1
1	0	0	0
0	1	1	0
1	1	0	1
1	0	1	1
1	1	1	1

Now consider the following:

$$Y = AC + B\overline{C} \tag{3-35}$$

The truth table for this equation is the same as that for equation 3-34, so the two equations must be equal, yielding the third identity:

$$AB + AC + B\overline{C} = AC + B\overline{C} \tag{3-36}$$

Boolean Expressions for Composite Gates

As discussed in the previous chapter, we are interested in three composite gates: NAND, NOR, and XOR. The Boolean expressions and identities already presented can be used to describe these composite gates.

The NAND gate (Figure 3.9) is an AND gate followed by a NOT gate. Its Boolean equation is thus:

$$Y = (\overline{AB}) \tag{3-37}$$

From DeMorgan's law (equation 3-22) it can be seen that the NAND gate is also described by the following:

$$Y = \overline{A} + \overline{B} \tag{3-38}$$

The NOR gate (Figure 3.10) is an OR gate followed by a NOT gate. Its Boolean equation is thus:

$$Y = (\overline{A + B}) \tag{3-39}$$

From DeMorgan's law (equation 3-21) it can be seen that the NOR gate is also described by the following:

$$Y = \overline{A}\,\overline{B} \tag{3-40}$$

The EXCLUSIVE-OR (XOR) gate (Figure 3.11) is complicated compared to the NAND and NOR gates, but its equation is straightforward:

$$Y = (A\overline{B}) + (\overline{A}B) \tag{3-41}$$

A comparison of equations 3-37 through 3-41 with the truth tables in Figures 3.9 through 3.11 shows that the Boolean equations yield the appropriate truth tables.

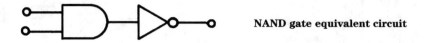

NAND gate equivalent circuit

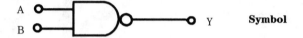

Symbol

Truth table

Inputs		Output
A	**B**	**Y**
0	0	1
0	1	1
1	0	1
1	1	0

Figure 3.9 The NAND gate.

NOR gate equivalent circuit

Symbol

Truth table

Inputs		Output
A	**B**	**Y**
0	0	1
0	1	0
1	0	0
1	1	0

Figure 3.10 The NOR gate.

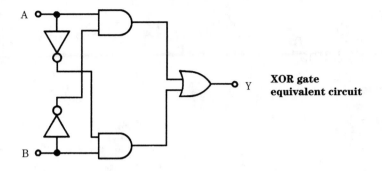

XOR gate
equivalent circuit

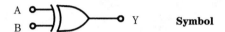   Y   **Symbol**

Truth table

Inputs		Output
A	**B**	**Y**
0	0	0
0	1	1
1	0	1
1	1	0

Figure 3.11 The EXCLUSIVE-OR (XOR) gate.

Summary

Boolean algebra provides a convenient shorthand for describing logic operations, particularly those operations of the logic gates discussed in the previous chapter. A summary of the laws and identities of Boolean algebra follows.

Operations

$$\times = \text{AND}$$
$$+ = \text{OR}$$
$$\overline{A} = \text{NOT } A$$

Laws

Commutative

$$A + B = B + A$$
$$AB = BA$$

Associative

$$A + (B + C) = (A + B) + C$$
$$A(BC) = (AB)C$$

Distributive

$$A(B + C) = AB + AC$$
$$A + BC = (A + B)(A + C)$$

Absorptive

$$A(A + B) = A$$

DeMorgan's

$$\overline{(A + B + C)} = \overline{A}\,\overline{B}\,\overline{C}$$
$$\overline{(ABC)} = \overline{A} + \overline{B} + \overline{C}$$

Identities

$$AA = A$$
$$X(1 + A + B + C + D \ldots) = X$$
$$\overline{\overline{A}} = A$$
$$\overline{\overline{(AB)}} = AB$$
$$\overline{\overline{(A + B)}} = A + B$$
$$AB + A\overline{B} = A$$
$$A + \overline{A}B = A + B$$
$$AB + AC + B\overline{C} = AC + B\overline{C}$$

4

Computer Arithmetic and Coding

The logic gates that form the heart of the programmable controller (PC) are binary devices, so some familiarity with the binary number system is required of PC users. Other number systems and codes are also used, including the octal and hexadecimal number systems, and the binary-coded decimal (BCD), Gray, Baudot, and ASCII codes. This chapter reviews these number systems and codes.

The Decimal Number System

The *decimal* number system is presented first because it is the one with which we are most familiar. In this numbering system, the *base*, or *radix*, is 10, which is also the number of digits in this system: 0 through 9. With these 10 digits, numbers of any magnitude can be generated as powers of 10.

To demonstrate this fact, we introduce *scientific notation*. Although scientific notation is generally used to represent numbers that are either very large or very small, in this case it is a convenient method of demonstrating the decimal number system. Scientific notation involves a base (or radix) and an exponent, or superscript. As mentioned before, the base is 10 for the decimal system. The exponent tells how many times the base is involved in self-multiplication. For example, 10^1 is just 10, while 10^2 is 10×10, or 100. Likewise, 10^3 is $10 \times 10 \times 10$, or 1000, and 10^4 is $10 \times 10 \times 10 \times 10$, or 10,000. A zero power of 10 also exists. Any number raised to the zero power is 1. Thus, $10^0 = 1$.

Consider the number 1234. This number is the summation of 1 thousand, 2 hundreds, 3 tens, and 4 ones. In other words

$$1234 = 1 \times 10^3 + 2 \times 10^2 + 3 \times 10^1 + 4 \times 10^0 \qquad (4\text{-}1)$$

Each *digit* in the number occupies a position that has a certain *weight*. The position farthest to the right (digit 4) has the weight of 10^0, or 1. The position second from the right (digit = 3) has the weight 10^1, or 10. This process continues to the left and can extend to infinity. Another example is presented in Figure 4.1.

As can be seen from Figure 4.2, decimal numbers can also be represented in this way. Positions to the right of the decimal place have negative exponential weights, such as 10^{-1} (0.1) and 10^{-2} (0.01).

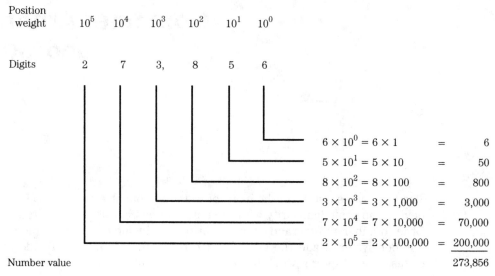

Figure 4.1 Representation of the decimal number 273,856.

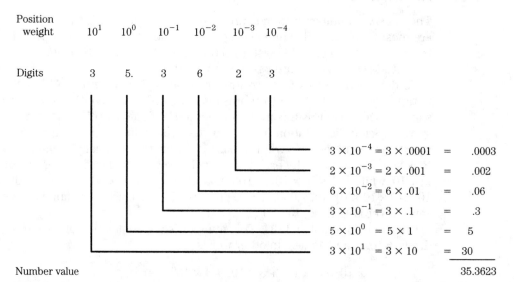

Figure 4.2 Representation of the decimal number 35.3623.

Because of our familiarity with the decimal number system, we do not stop to consider the weight of each position. We automatically recognize that the number 432 represents 4 hundreds, 3 tens, and 2 ones. When using other numbering systems (i.e., systems with bases other than 10), however, it is necessary to evaluate numbers using the method shown in Figure 4.1. The next section illustrates this concept.

The Binary Number System

The binary number system has a base of 2, which is also the number of digits in the system: 0 and 1. With these two digits, numbers of any magnitude can be generated as powers of 2.

Each digit in any number occupies a position with a certain weight. This concept can be demonstrated using Figure 4.3, which shows the binary number 11011. The position farthest to the right (digit 1) has the weight 2^0, or 1. (Remember that any number raised to the power of zero is 1.) The second position from the right (digit 1) has the weight 2^1, or 2. The third position from the right (digit 0) has the weight 2^2, or 4. The fourth position from the right (digit 1) has the weight 2^3, or 8. The furthermost left position (digit 1) has the weight 2^4, or 16. Thus, the binary number 11011 is equivalent to the decimal number 27:

$$11011 = 1 \times 2^4 + 1 \times 2^3 + 0 \times 2^2 + 1 \times 2^1 + 1 \times 2^0$$
$$11011 = 1 \times 16 + 1 \times 8 + 0 \times 4 + 1 \times 2 + 1 \times 1 \qquad (4\text{-}2)$$
$$11011 = 16 + 8 + 2 + 1 = 27$$

The advantages of using the binary number system should be obvious. Logic gates exist in one of two possible states: on or off, 1 or 0. A number system that has only two digits is natural for digital logic operations. Thus, the binary number system is the primary language of logic circuits, computers, and programmable controllers.

Converting decimal numbers to binary numbers

It is easy to convert a number from the binary to the decimal number system. You simply use the method outlined above to obtain equation 4-2. In other words, each digit in the binary number is multiplied by its weighting factor, and the total is summed. Converting a number from the decimal to the binary number system is not quite as straightforward, but it is not terribly complicated. The method for doing this is often called the *dibble-dabble method*; it is demonstrated in Figure 4.4.

In this figure, we are converting the decimal number 79 to its binary equivalent. The first step involves dividing 79 by 2:

79/2 = 39, with a remainder of 1

The remainder of 1 becomes the first digit in the binary number, i.e., the digit with a weight of 2^0, or 1. The second digit is obtained by dividing 39 by 2:

39/2 = 19, with a remainder of 1

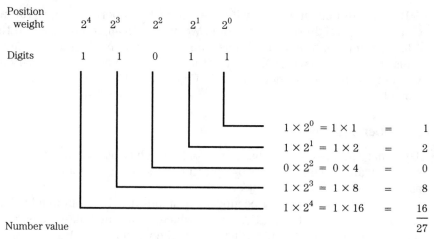

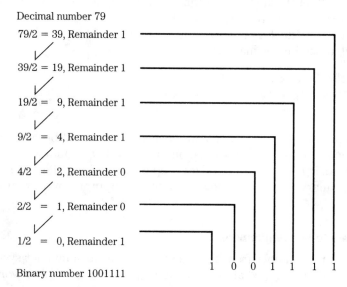

Figure 4.3 Representation of the binary number 11011.

Figure 4.4 The dibble-dabble method for converting decimal numbers to binary numbers.

The remainder of 1 becomes the second digit in the binary number, or the digit with a weight of 2^1, or 2. We continue this process:

79/2 = 39, remainder 1
39/2 = 19, remainder 1
19/2 = 9, remainder 1
9/2 = 4, remainder 1
4/2 = 2, remainder 0
2/2 = 1, remainder 0
1/2 = 0, remainder 1

When read from bottom to top, the remainders give us the binary equivalent of decimal 79, or binary 1001111.

It is easy to check this answer. Simply convert the binary number 1001111 to decimal, and check that the answer is 79:

$$1001111 = 1 \times 2^6 + 0 \times 2^5 + 0 \times 2^4 + 1 \times 2^3 + 1 \times 2^2 + 1 \times 2^1 + 1 \times 2^0 \qquad (4\text{-}3)$$
$$1001111 = 1 \times 64 + 0 \times 32 + 0 \times 16 + 1 \times 8 + 1 \times 4 + 1 \times 2 + 1 \times 1$$
$$1001111 = 64 + 8 + 4 + 2 + 1$$
$$1001111 = 79$$

The dibble-dabble method is thus an effective method of converting a number from the decimal number system to the binary number system.

Binary arithmetic

We are all familiar with the four basic arithmetical operations in the decimal number system: addition, subtraction, multiplication, and division. These operations can also be performed in the binary number system.

The addition table for the binary number system is quite simple:

$$0 + 0 = \quad 0 \qquad\qquad (4\text{-}4)$$

$$0 + 1 = \quad 1 \qquad\qquad (4\text{-}5)$$

$$1 + 1 = \quad 10 \qquad\qquad (4\text{-}6)$$

$$10 + 1 = 11 \qquad\qquad (4\text{-}7)$$

To see how this table is used in the addition of binary numbers, let us add two numbers:

```
  1011
+ 1100
```

We begin, as in the decimal number system, with the column of numbers farthest to the right (i.e., the *least significant digit*) and proceed to the left-most column of numbers (i.e., the *most significant digit*). Equation 4-5 allows us to fill in the three columns to the right, while equation 4-6 allows us to fill in the column of most significant digits:

```
  1011
+ 1100
------
 10111
```

The result can be checked by converting all the numbers to decimal. Binary number 1011 is decimal number 11. Binary number 1100 is decimal number 12. Their sum is 23. In the binary number system, 23 is represented as 10111. Thus, the result checks out.

Numbers are "carried over" to the next column in binary addition just as in decimal addition. Addition of the binary numbers 10111 and 11111 provides a convenient example. Beginning with the least significant digits, $1 + 1 = 10$, we record the 0 and carry over the 1. The next column is $1 + 1 = 10$, plus the carried-over 1, which yields a total of 11. We record the least significant 1 and carry over a 1. This system continues to the column of most significant digits:

```
  10111
+ 11111
--------
 110110
```

To check this solution, convert all numbers to decimal. Binary 10111 equals decimal 23, and binary 11111 equals decimal 31. Their sum is decimal 54, which is binary 110110. The solution thus checks out.

Binary addition is easily accomplished with logic circuits, as seen in the next section. The rules of subtraction in the binary number system are as follows:

$$0 - 0 = 0 \tag{4-8}$$

$$1 - 0 = 1 \tag{4-9}$$

$$1 - 1 = 0 \tag{4-10}$$

$$10 - 1 = 1 \tag{4-11}$$

As an example, consider subtracting 10 from 11:

```
  11
- 10
-----
```

Equation 4-9 gives us the solution for the least significant digit, while equation 4-10 gives us the solution for the most significant digit:

```
  11 (minuend)
- 10 (subtrahend)
-----------------
   1 (difference)
```

As in the decimal system, the zero in the most-significant-digit column is not written. As a check, binary 11 equals decimal 3, binary 10 equals decimal 2, and their difference is 1, in either number system.

Because $0 - 1$ is not allowed in binary subtraction, when 0 appears in the minuend above a subtrahend digit of 1, we must "borrow" from the next digit to the left, just as in decimal subtraction. An example should make this clear. Assume that we wish to subtract 489 from 537 in the decimal number system:

```
column    (a) (b)(c)
           5   3  7
          -4   8  9
          -----------
```

We begin with column c. Obviously, 7 is less than 9, so the 7 "borrows" a 1 (actually a 10) from column b:

```
column     (a) (b) (c)
            5   3  (17)
           -4   8   9
           _____
                    8
```

The subtraction, $17 - 9$, is performed, and we are finished with column c.

In column b, the 1 borrowed by column c must be repaid, but it is *repaid to the subtrahend*. Thus, the 8 in the subtrahend gains a 1 and becomes a 9:

```
column    (a)  (b)   (c)
           5    3   (17)
          -4   (9)    9
          _____
                       8
```

Now 3 is less than 9, so it borrows a 1 from column a.

```
column    (a)   (b)    (c)
           5   (13)   (17)
          -4   ( 9)     9
          _____
                 4       8
```

The subtraction $13 - 9$ completes our action with column b. We move to column c and repay the borrowed 1 to the 4 in the subtrahend:

```
column    (a)   (b)    (c)
           5   (13)   (17)
         -(5)  -(9)     9
         _____
           0     4       8
```

We drop the left-most zero (column a), and note that the result of $537 - 489$ is 48. Adding the difference, 48, to the subtrahend, 489, results in 537, which is the minuend. Thus, the result checks out.

The same procedure is followed in binary subtraction. Consider the problem of 11010 minus 1111:

```
column    (a)  (b)  (c)  (d)  (e)
           1    1    0    1    0
         -       1    1    1    1
         _____
```

We begin with column e. In column e the minuend borrows a 1 from column d:

```
column    (a)  (b)  (c)  (d)  (e)
           1    1    0    1   (10)
         -       1    1    1    1
         _____
                                 1
```

We use equation 4-11 for column e. This 1 gets repaid to the subtrahend in column d.

column	(a)	(b)	(c)	(d)	(e)
	1	1	0	1	(10)
−		1	1	(10)	1
					1

Column d now borrows a 1

column	(a)	(b)	(c)	(d)	(e)
	1	1	0	(11)	(10)
−		1	1	(10)	1
				1	1

which is paid back to the subtrahend in column c:

column	(a)	(b)	(c)	(d)	(e)
	1	1	0	(11)	(10)
−		1	(10)	(10)	1
				1	1

The minuend in column c now borrows a 1:

column	(a)	(b)	(c)	(d)	(e)
	1	1	(10)	(11)	(10)
−		1	(10)	(10)	1
			0	1	1

which gets paid back to column b of the subtrahend:

column	(a)	(b)	(c)	(d)	(e)
	1	1	(10)	(11)	(10)
−		(10)	(10)	(10)	1
		0	1	1

The process is repeated to completion:

column	(a)	(b)	(c)	(d)	(e)
	1	(11)	(10)	(11)	(10)
− (1)	(10)	(10)	(10)	1
	0	1	0	1	1

Dropping the left-most zero yields the result:

```
  11010
 – 1111
 ──────
   1011
```

This result can be checked in two ways. First, add the difference to the subtrahend. The result should be the minuend.

```
   1011
 + 1111
 ──────
  11010
```

Second, convert the numbers to decimal and perform the subtraction again:

```
  11010 =   26
 – 1111 = – 15
 ──────────────
   1011 =   11
```

The subtraction checks both ways.

In the decimal number system, subtraction can be viewed as the addition of a negative number to a positive number. Logic circuits, computers, and PCs do not recognize negative numbers; thus, subtractions performed by electronic logic devices cannot be carried out using the method described above. The method of *complements* is used instead by both computers and PCs to perform binary subtraction.

Two types of complements exist: the *one's complement* and the *two's complement*. Consider the one's complement first. In the one's complement method, an extra digit is placed in the left-most column of the number, or the most-significant-digit column. This digit is 0 if the number is positive, and 1 if the number is negative. If the number is negative, each digit in the number is inverted (i.e., 1s become 0s, and 0s become 1s). For example, the binary number 1011 is equal to the decimal number 11. We now give the number a *sign* digit: 0 is positive, 1 is negative: binary 01011 = decimal +11. To represent the decimal number –11 in the binary system, we invert all the digits: binary 10100 = decimal –11.

What is the result when we add decimal numbers +11 and –11? The result is 0, of course. In the binary number system, however, we have

```
   01011
 + 10100
 ──────
   11111
```

which is not zero. To obtain zero, we must add a 1 to the least significant digit and ignore the most significant digit (MSD):

```
   01011
 + 10100
 +     1
 ───────
 100000 = 0, when MSD is ignored.
```

This example outlines how subtraction is performed with the one's complement method: find the complement of the subtrahend, add 1 to it, add the result to the minuend, and ignore the most significant digit.

For a second example, let us subtract decimal 22 from decimal 41 (result = 19) in the binary system.

```
decimal     binary
  41         101001
 –22        – 10110
 ───        ───────
  19          10011
```

To begin, add the positive-sign digit (0) to 101001 and take the complement of 10110, making sure to add the negative sign (1) as the most significant digit.

```
  0101001
+  101001
─────────
```

Next, add 1 to the subtrahend, and then add the subtrahend to the minuend:

```
  0101001
+  101010
─────────
  1010011
```

If we ignore the most significant digit (the 1 farthest to the left), the result is 010011, or positive 10011. Thus, the solution checks out.

As a final example, let's subtract 1011 (decimal 11) from 1110 (decimal 14). The result should be 11 (decimal 3);

```
   1110          01110          01110
 – 1011    →   + 10100    →   + 10101
 ──────        ───────        ───────
                              100011 =
 +11
```

The solution is thus correct.

The two's complement method is somewhat different from the one's complement. The extra digit in the most significant column is still used (0 = positive and 1 = negative), and some digits are inverted. Inversion (from right to left) in the two's complement, however, occurs *only* after the first 1 is detected.

An example should make this rule clear. Binary number +1110 (01110, or decimal +14) has the one's complement of 10001:

binary number   01110
one's complement   10001

To find the two's complement, we begin with the least significant digit, which is 0. Because a 1 has not been detected, the 0 is left unchanged. We move left to the next column. The digit is a 1, which is the first detected. It is left unchanged, *but all digits after this first 1 are then inverted*:

binary number   01110
two's complement   10010

Thus, in two's complement terminology, 10010 is –14.

When the original binary number is added to its two's complement, the result should be 0. As seen below, it is 0 *if the most significant digit is ignored*.

```
   01110
 + 10010
 ───────
  100000
```

This example provides us with the rules of subtraction using two's complements. First, find the two's complement of the subtrahend. Second, add it to the minuend. Third, ignore the most significant digit of the sum. The following two examples demonstrate these rules.

decimal	binary	two's complement
41	101001	0101001
– 22	– 10110	101010
19	10011	1010011

or +10011.

decimal	binary	two's complement
14	1110	01110
– 11	– 1011	10101
3	11	100011

or +11.

The only difference between subtraction with two's complements and subtraction with one's complements is that, with two's complements, it is not necessary to add a 1 to the complement.

Once addition and subtraction are mastered, multiplication and division are performed easily. Multiplication is just multiple addition. For example, 55 × 12 is nothing more than 55 added to itself twelve times. Division is multiple subtraction. For example, to divide 16 by 8, we subtract 8 as follows:

$$
\begin{array}{r}
16 \\
-\ 8 \text{ first subtraction} \\
\hline
8 \\
-\ 8 \text{ second subtraction} \\
\hline
0
\end{array}
$$

yielding a total of two subtractions; thus, 16/8 = 2.

Of course, the same holds true for the binary number system. 1101 × 101 is just 1101 added to itself 101 times:

$$
\begin{array}{r}
1101 \\
\times\ 101 \\
\hline
1101 \\
1101 \\
\hline
1000001
\end{array}
$$

In the same way, 1100 divided by 110 is just the number of times that 110 can be subtracted from 1100.

$$
\begin{array}{r}
10 \\
110\overline{)1100} \\
\underline{110} \\
0
\end{array}
$$

Using complements, subtraction can be performed as addition. Multiplication and division are nothing more than multiple additions and subtractions, respectively. Thus, all four functions (addition, subtraction, multiplication, and division) can be performed in a programmable controller by an adding circuit, which is discussed in the next section.

Binary adder circuits

The two basic types of logic circuit adders are the *half-adder* and the *full adder*. The half-adder symbol, circuit, and truth table are shown in Figure 4.5. The half-adder adds two digits. It accepts two inputs (the digits to be added) and yields two outputs: the sum, or S, output and the carry, or C, output. The circuit consists of an EXCLUSIVE-OR (XOR) gate and an AND gate. (Recall from chapter 2 that the XOR gate is a composite gate, comprising two NOT, two AND, and one OR gate, as shown in Figure 2.14.) The sum output is given by equation 4-12:

$$S = A\overline{B} + \overline{A}B \tag{4-12}$$

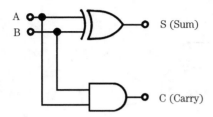

Truth table

Inputs		Outputs	
A	**B**	**S**	**C**
0	0	0	0
0	1	1	0
1	0	1	0
1	1	0	1

Figure 4.5 Symbol, circuit, and truth table for the half-adder.

while the carry output is given by equation 4-13:

$$C = AB \tag{4-13}$$

In adding two digits with the half-adder, the operation is performed as follows:

$$\begin{array}{r} A \\ + B \\ \hline CS \end{array}$$

With this in mind, the truth table in Figure 4.5 yields the following results:

$$\begin{array}{rrrr} 0 & 0 & 1 & 1 \\ +0 & +1 & +0 & +1 \\ \hline 00 & 01 & 01 & 10 \end{array}$$

The full adder adds three digits: A, B, and a carry input CI from a previous full adder or half-adder. This relationship is shown in Figure 4.6. The logic equations for the sum and carry outputs are the following:

$$S = A\overline{B}\,\overline{CI} + \overline{A}B\overline{CI} + \overline{A}\,\overline{B}CI + ABCI \tag{4-14}$$

$$C = AB + (A + B)CI \tag{4-15}$$

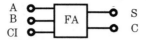

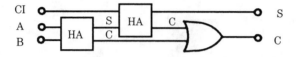

Truth table

Inputs			Outputs	
A	**B**	**CI**	**S**	**C**
0	0	0	0	0
0	0	1	1	0
0	1	0	1	0
1	0	0	1	0
0	1	1	0	1
1	1	0	0	1
1	0	1	0	1
1	1	1	1	1

Figure 4.6 Symbol, circuit, and truth table for the full adder.

Using both the full adder and half-adder, addition can be performed on numbers containing many digits. A simple example of 11 + 10 is shown in Figure 4.7.

Adder circuits can be quite complex. This book does not explore the design and construction of adder circuits, but from the examples given above, you can see that logic gates can be combined to perform addition, and that addition can be used to perform the other three arithmetical operations: subtraction, multiplication, and division. Multitudes of logic gates can be fabricated onto one small integrated circuit chip in such a way as to perform all these operations. Such a chip is called a *central processing unit* (CPU), which is the heart of a computer or programmable controller. The CPU is thus nothing more than a collection of logic gates.

Before proceeding to the next section, we introduce several definitions. A *bit* is a digit in the binary number system. The binary number 1011 has four bits, while the binary number 10101010 has eight bits. A *nibble* is a group of four bits. A *byte* is a group of eight bits. A *word* is a group of one or more bytes. Using this terminology, the most significant digit and least significant digit in the binary number system become the *most significant bit* and *least significant bit*, respectively.

The Octal Number System

Because logic circuits, computers, and PCs are binary devices, their primary language is the binary number system. Other number systems, however, are usually employed for communication with the computer (such as input/output addressing) and for data storage (such as memory). The two principal number systems of communication are the *octal number system* and the *hexadecimal number system*. The octal number system is discussed in this section.

The base, or radix, of the octal number system is 8, which is also the number of digits in the system: 0 through 7. (Note that 8 is also the number of bits in a byte.) As before, numbers greater than 7 are represented as powers of 8:

$$\text{octal } 1234 = 1 \times 8^3 + 2 \times 8^2 + 3 \times 8^1 + 4 \times 8^0 \qquad (4\text{-}16)$$
$$= 1 \times 512 + 2 \times 64 + 3 \times 8 + 4 \times 1$$
$$= 512 + 128 + 24 + 4$$
$$= 668 \text{ (decimal)}$$

The advantage of the octal system over the binary system is immediately apparent. Four digits are required to represent the decimal 668 in the octal system. In the binary system, 10 bits are required:

binary 1010011100 = decimal 668

The base of the octal system is 8, which is 2^3, providing a convenient method of converting between octal and binary systems. To convert from binary to octal, we group the binary number into three-bit groups, beginning with the least significant bit. We then convert into octal by simply converting the groups into decimal. This procedure is shown in Figure 4.8. To convert from octal to binary, convert each octal digit into a three-bit binary number, beginning with the least significant digit, as shown in Figure 4.9.

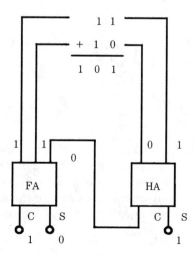

Figure 4.7 Addition of 11 + 10 using one full adder and one half-adder.

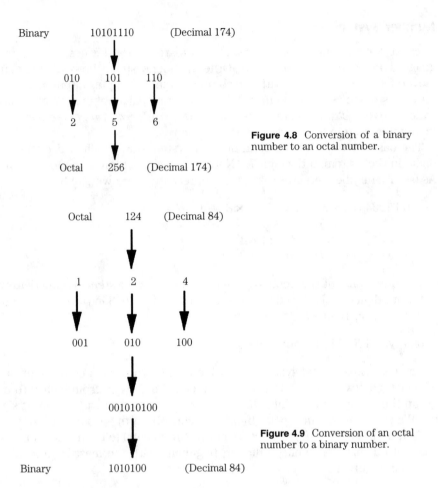

Binary   10101110   (Decimal 174)

010   101   110

2   5   6

Octal   256   (Decimal 174)

Figure 4.8 Conversion of a binary number to an octal number.

Octal   124   (Decimal 84)

1   2   4

001   010   100

001010100

Figure 4.9 Conversion of an octal number to a binary number.

Binary   1010100   (Decimal 84)

The easiest way to convert from decimal to octal, and vice versa, is to first convert the number to binary and then use the method outlined in the preceding paragraph.

The Hexadecimal Number System

The base, or radix, of the hexadecimal number system is 16, which is also the number of characters in the system: 0 through 9 and A through F. A through F represent numbers 10 through 15 because these numbers have two digits each. The use of two digit numbers would not only be confusing, but would also increase the number of bits necessary to represent bytes or words. Numbers greater than F are represented as powers of 16:

hexadecimal $B13 = 11 \times 16^2 + 1 \times 16^1 + 3 \times 16^0$
$$= 11 \times 256 + 1 \times 16 + 3 \times 1 \qquad (4\text{-}17)$$
$$= 2816 + 16 + 3$$
$$= 2835 \text{ (decimal)}$$

The preceding example displays the advantage of using the hexadecimal number system. A number that requires four decimal digits can be represented with only three hexadecimal characters.

Because $2^4 = 16$, each character of a hexadecimal number represents four binary bits, providing a convenient method for converting back and forth between the binary and hexadecimal number systems, as shown in Figures 4.10 and 4.11. To convert between hexadecimal and octal, or hexadecimal and decimal, use the binary number system as an intermediary.

The four basic number systems are compared in Table 4.1.

TABLE 4.1

Decimal	Binary	Octal	Hexadecimal
0	0	0	0
1	1	1	1
2	10	2	2
3	11	3	3
4	100	4	4
5	101	5	5
6	110	6	6
7	111	7	7
8	1000	10	8
9	1001	11	9
10	1010	12	A
11	1011	13	B
12	1100	14	C
13	1101	15	D
14	1110	16	E
15	1111	17	F

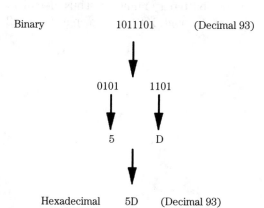

Figure 4.10 Conversion of a binary number to a hexadecimal number.

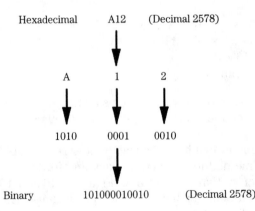

Figure 4.11 Conversion of a hexadecimal number to a binary number.

Binary Codes

A code is just a systematic method for presenting or transmitting information. The binary codes presented in the remainder of this chapter are designed to allow machines to communicate with one another; specifically, to allow the computer to communicate with external devices, such as input/output devices, also known as *peripherals*.

Obviously, one of the most important communications links is that of communication between the operator (or programmer) and the PC. The codes that follow translate the programmer's language into a format the PC or computer can use. This translation is done by assigning a unique combination of binary bits to each number, letter, or symbol. Of the codes that follow, the binary coded decimal, Gray, and ASCII codes are quite common. The Baudot code has fallen into disuse, primarily because of the superiority of ASCII. The ASCII code is a popular keyboard code for computer and PC terminals. The Baudot code was used primarily with teleprinter machines.

The binary coded decimal code

The great advantage of the binary coded decimal (BCD) code is its simplicity. The BCD code converts decimal numbers to four-bit binary numbers. Thus, decimal numbers 0 through 9 are represented by BCD 0000 through 1001, as shown in Table 4.2.

TABLE 4.2

Decimal	BCD
0	0000
1	0001
2	0010
3	0011
4	0100
5	0101
6	0110
7	0111
8	1000
9	1001

Conversion between decimal and BCD is easier than between decimal and binary. For example, decimal 23 can be represented in binary as 10111. To convert, one must recall the powers of 2 up to the fourth power. The number 23 in BCD is 0010 0011, where the first four bits represent the 2 and the second four bits represent the 3. One need only recall the powers of 2 up to the third power. Unfortunately, BCD requires more binary bits than the binary system to represent the same number (in this example, 8 in BCD versus 5 in binary).

The largest number that can be represented with four binary bits is 1111 (decimal 15). The largest required for BCD is 1001 (decimal 9). Thus, the numbers 1010, 1011, 1100, 1101, 1110, and 1111 are not used in BCD. These numbers are not wasted, however, as they can be used for sending instructions. For example, 1110 could be sent as an instruction to a stamping machine that tells the machine to change the die pattern.

An excellent example of an application of the BCD code is the pocket calculator. Each number key, when pushed, transmits a BCD signal to the calculator's CPU. For example, when the number 7 key is pressed, 0111 is sent to the calculator's internal digital circuits. Likewise, when the answer is determined and transmitted to the calculator's display, the BCD is converted to decimal for use by the seven-segment LED display.

The Gray code

In BCD, decimal 1 is represented as 0001, while decimal 2 is represented as 0010. To make the ascent from decimal 1 to decimal 2, two bits must change: the least significant bit from 1 to 0, and the bit to the left from 0 to 1. The Gray code (also known as cyclic code, or reflected binary code) requires only one bit to change between successive numbers. The Gray code is presented in Table 4.3.

TABLE 4.3

Decimal	Gray code
0	0000
1	0001
2	0011
3	0010
4	0110
5	0111
6	0101
7	0100
8	1100
9	1101
10	1111
11	1110
12	1010
13	1011
14	1001
15	1000

The Gray code is not suitable for arithmetic operations. Its primary use is in mechanical-to-electrical conversions, such as positioning applications.

The Baudot code

The Baudot code found its greatest application in teleprinter machines where, in addition to 10 digits, 26 letters of the alphabet and numerous symbols (such as "#") needed to be represented.

The Baudot code is a five-level code, which means that each character on the keyboard (i.e., each character of transmitted data) is represented by five binary bits. For example, when the letter F on the teleprinter machine is pressed, the machine transmits the binary code 10110. By using a shift key, the 64 keyboard characters could be represented with the 32 Baudot numbers (00000 through 11111).

Two principal drawbacks existed with the Baudot code: it contained only 32 numbers, and these numbers were not in ascending binary order. The ASCII code, described in the next section, effectively eliminated these problems.

The ASCII code

ASCII (pronounced "as-key") is an acronym for *American Standard Code for Information Interchange*. With this code, six, seven, or eight bits per character are used. With six bits per character, 2^6 or 64 characters can be represented. With seven bits, 2^7 or 128 characters can be represented. Thus, upper- and lowercase letters and special characters can be represented. ASCII also allows for control characters, which are used in communication. The seven-bit ASCII code is most common. The eight-bit code ($2^8 = 256$) is used for error-checking.

The ASCII code is an excellent keyboard code and is widely used in computer and PC keyboards. The keyboard translates the message from the high-level programming language (such as BASIC) into ASCII when sending it to the computer, and from ASCII back to the programming language when receiving a message from the computer. The ASCII code is provided in appendix D.

Summary

The basic number systems of PCs (binary, octal, and hexadecimal) have been presented, as well as several codes (BCD, Gray, Baudot, and ASCII). These systems and codes are used for PC operations, including communication between operator and the PC and between the PC and its peripherals.

5

Relays in
Control Applications

Chapter 1 stated that programmable controllers were developed to replace relays in control applications. Chapter 2 discussed the use of relays to perform the functions of logic gates. In this chapter, we examine the use of relays in control applications and introduce *relay-ladder diagrams*. Relay-ladder programming is then explored in chapter 9.

The Relay

A relay is nothing more than an electromagnetic switch. An electromagnet is used in a relay to activate or deactivate the switch, as shown in Figure 5.1. When current flows through the coil, it generates a magnetic field, which attracts the armature. The moving contact on the armature is drawn down to the stationary, normally open contact and completes the circuit. Thus, current flows to the lamp, just as if a switch in the line had been closed. When current flow to the coil is interrupted, its magnetic field collapses, the armature is no longer attracted to the coil, and the lamp circuit is opened.

The name *relay* comes from the history of the device, when it was first used in telegraphy. In the early days of the telegraph, dc voltage was used. The difficulty with dc voltage was that its effective range was limited to less than 20 miles; telegraph signals were so weakened after transmission of 20 miles that they were unintelligible. To avoid erecting telegraph stations every 20 miles and using human operators to relay signals, the relay was invented. Relays were placed along the telegraph lines at about 20 miles distance. They detected weak telegraph signals, and "relayed" them with increased strength to stations further down the line, as illustrated in Figure 5.2.

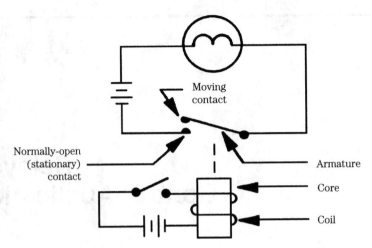

Figure 5.1 The relay.

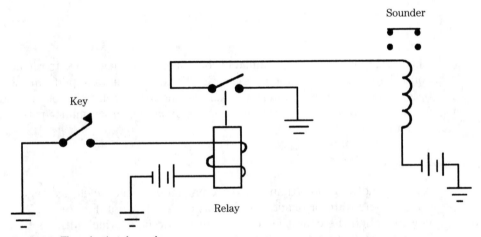

Figure 5.2 The relay in telegraphy.

The use of ac voltage in telegraphy greatly increased the range between relays, but relays are still used in telegraphy. Relays are also used in a variety of other applications. The typical automobile provides an example of why a relay could be used instead of a switch. A starter in an automobile draws a large current of up to several amps when it is turned on. If a switch were used to carry that much current, it would need to be very heavy and would be difficult to throw. A relay placed in the circuit, however, requires only a very small current to energize the relay coil, and the switch to the relay then becomes smaller and easier to throw. The high current requirement of the starter motor is met using a relay with large, high-current contacts.

The simple relay shown in Figure 5.1 is a single-pole single-throw (SPST) relay. It closes (or opens) a single set of contacts. Like switches, relays can be purchased with a variety of poles and throws. Figure 5.3 shows two examples. In Figure 5.3(a) we have a single-pole double-throw relay, which has two sets of stationary contacts: those that are normally open, and those that are normally closed. *Normally* refers to the contact condition when the relay coil is not energized. Thus, a normally open contact does not touch the moving contact when the coil is not energized. Normally open and normally closed are typically abbreviated N.O. and N.C., respectively. In Figure 5.3(b), a three-pole double-throw relay is shown. This relay is essentially three single-pole double-throw relays, all activated by the same relay coil.

Industrial Relays

Relays used in industrial control applications operate on the same general principles as the relays described in the preceding section. There are, however, a few key differences. Industrial control relays are extremely rugged; they are built to survive in the sometimes rigorous manufacturing environment. Note that this difference is similar to one of the key differences between the personal computer and the programmable controller. In addition industrial control relays almost always provide both N.O. and N.C. contacts; they are almost always double-throw relays. Finally, armature and moving contacts are different from those relays just described.

Figure 5.4 shows that the armature moves the moving contacts up or down. The moving contacts are now paired; there are two of them. This pair helps to dissipate the transient voltages generated by the arc at the contact point more quickly than a single moving contact.

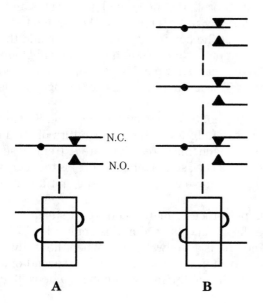

Figure 5.3 Other relays. A) Single-pole double-throw (SPDT) relay.
B) Three-pole double-throw relay.

A   B

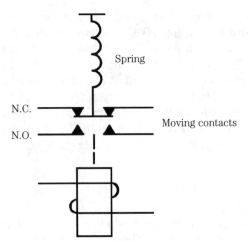

Figure 5.4 Industrial relay construction.

Relay Control Examples

Relays exist in one of two states: on or off. Thus, relays are on-off controllers. This section presents two examples of how relays are used in on-off control applications.

The first example is very simple and involves a weighing platform at a shipping station (Figure 5.5). As filled, sealed cartons proceed down a conveyor belt, they pass over a pressure plate. If the carton is completely filled, the pressure plate activates a switch that energizes an SPDT relay. The normally open contacts of the relay are connected to a green lamp, while the normally closed contacts are connected to a red lamp. Thus, if the box is filled properly, the green light appears and the shipping clerk knows that the box is ready to be shipped. If the red light stays illuminated, the shipping clerk knows that the box is not properly filled, and the box can be set aside to be refilled before shipping. Note that although the refinements shown in Figure 5.4 are not included in the relay shown in Figure 5.5, it is assumed that the relay is an industrial control relay. This assumption is made throughout the remainder of this book.

The second example involves the control of steam temperature. Steam temperature is typically controlled with water. To lower the temperature of steam, one injects water into the steam line. Many chemical reactions are extremely sensitive to temperature. If reaction temperature (i.e., steam temperature) is too high, reactions can become explosive. Thus, preventing excessive high temperatures is an important control function.

Figure 5.6 shows a very simple process for the control of steam temperature. (Note that a single-pole single-throw relay is used in this application.) Two objects are placed in the steam line. The first is a feedwater line. When the feedwater pump is turned on, cooling water is sprayed into the steam line. The second object in the steam line is a *thermocouple*, which is a device that produces a dc voltage proportional to temperature.

The control system operates as follows. The thermocouple produces a positive dc voltage proportional to temperature. Its signal is fed into the inverting input of an operational amplifier. A positive reference voltage, equal to the temperature at which the relay should be activated, is fed into the noninverting input of the operational amplifier. The operational amplifier in this mode acts as a *comparator*. If the reference voltage (critical temperature) is greater than the thermocouple voltage (actual temperature), the output from the comparator is positive, which reverse-biases the diode. No current can then flow to the relay. If, however, the thermocouple voltage exceeds the reference voltage (which means the critical temperature is exceeded), the output of the comparator becomes negative, which forward-biases the diode and triggers the relay.

When the relay is triggered, an ac voltage is applied to a constant-speed pump that sprays cooling water into the steam line, reducing the temperature of the steam and the voltage from the thermocouple. When the thermocouple voltage falls below the reference voltage (the actual temperature falls below the critical temperature), the relay is deenergized, and the pump turns off.

The area of Figure 5.6 enclosed by dashed lines can be thought of as a *temperature-activated switch* that is normally open. At a certain temperature, the switch closes, allowing current flow to the coil of the relay. Note that this relay-control scheme offers only on-off control; it does not allow incremental control of steam temperature.

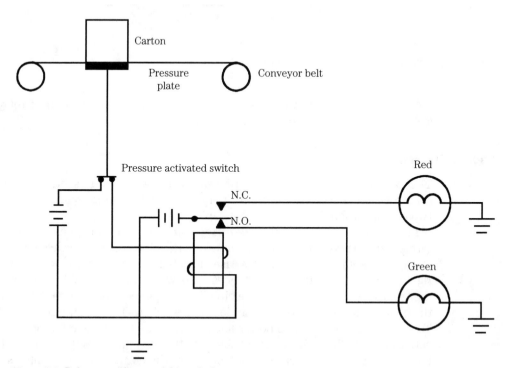

Figure 5.5 Relay control at a weighing platform.

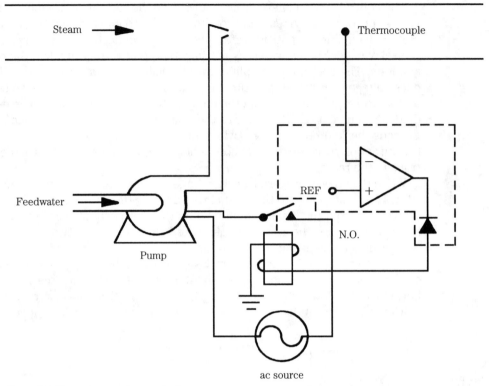

Figure 5.6 Steam temperature control.

In the next section, a convenient shorthand for the description of relay-control systems (such as the two examples given above) is presented.

Ladder Diagrams

The two relay-control examples presented in the previous section can be represented by a convenient shorthand notation called the *ladder diagram*. The next two examples should clarify the concept of the ladder diagram.

Figure 5.7 shows the *relay-ladder diagram* for the control circuit in Figure 5.5. The term *ladder* derives from the appearance of the diagram. The vertical lines in the diagram represent power lines; in this case, the left line is at a positive dc potential and the right line is negative. In many instances, ac power lines are used. Power flows through the "rungs" of the ladder when the circuit is completed. For example, in rung 1, when the pressure-activated switch PS is closed, power flows through the coil R1 of the relay. To the right of the negative power line at rung 1 are the cross-referencing comments N.C. 2 and N.O. 3. These comments indicate the rung numbers that represent the contacts of relay R1. In this case, rung 2 describes the normally closed contacts of the relay, while rung 3 describes the normally open contacts of the relay.

Rung 2 in Figure 5.7 shows the symbol for a normally closed contact and a red lamp, while rung 3 shows the symbol for a normally open contact and a green lamp. The R1 above both sets of contacts simply reminds us that these are contacts for relay number 1, whose coil circuit is shown in rung 1.

The three rungs in Figure 5.7 completely describe the control system of Figure 5.5. When switch PS is closed, the relay coil is energized and power flows through rung 3, illuminating the green lamp. When PS is open, the relay coil is not energized, and power flows through rung 2, illuminating the red lamp.

Figure 5.8 shows a second example of a relay ladder diagram. This figure describes the control circuit shown in Figure 5.6, the circuit that controls steam temperature. The circuitry enclosed in dashed lines in Figure 5.6 is represented in Figure 5.8 as a normally open temperature-activated switch (TAS) in rung 1 of the ladder diagram. When the temperature-activated switch is closed (the steam temperature exceeds a critical temperature), the coil of the relay R1 is energized.

Because the relay is an SPST relay, only one additional rung is needed to complete the relay ladder diagram. This appears as rung 2 in Figure 5.8. When the relay coil is energized, the normally open contacts of the relay are closed, and power is supplied to the pump motor (MTR in Figure 5.8). Thus, cooling water is supplied to the steam line until the TAS in rung 1 opens again.

While the relay ladder diagrams are descriptive, they are often replaced with *contact-ladder diagrams*. Contact-ladder diagrams show only contacts and outputs, which are indicated by parentheses. Contact-ladder diagrams for Figures 5.7 and 5.8 are shown in Figures 5.9 and 5.10, respectively.

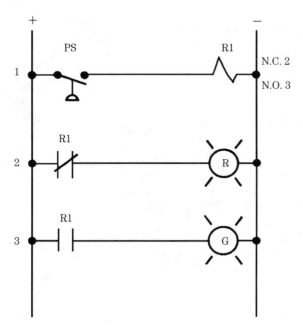

Figure 5.7 Relay ladder diagram for Figure 5.5.

Summary

This chapter presented two examples of relays used in control applications. Relay and contact ladder diagrams have been introduced. Chapter 9 further develops relay and contact ladder diagrams and their use in programming PCs.

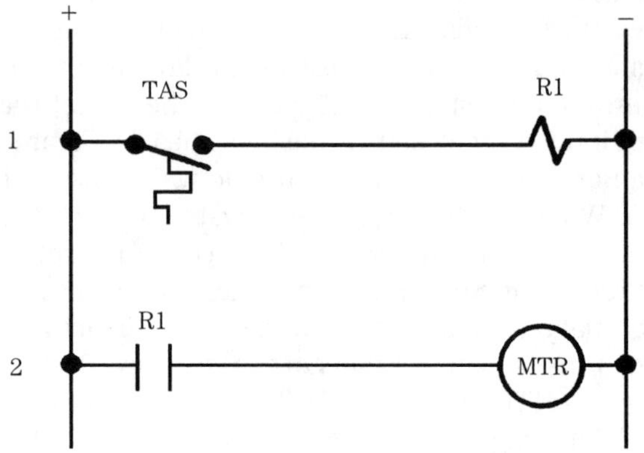

Figure 5.8 Relay ladder diagram for Figure 5.6.

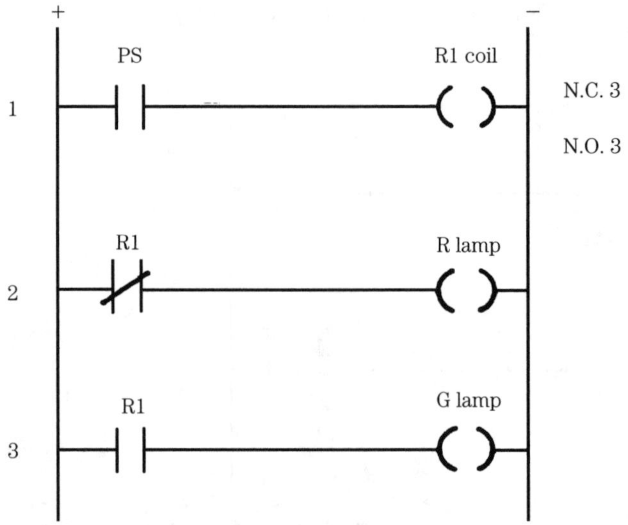

Figure 5.9 Contact ladder diagram for Figure 5.7.

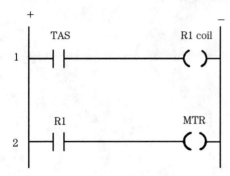

Figure 5.10 Contact ladder diagram for Figure 5.8.

6

The Central Processing Unit

The preceding five chapters presented basic information necessary for an understanding of programmable controllers and their operation. In this chapter, and the two that follow, we begin an examination of the hardware associated with programmable controllers. This chapter examines the central processing unit (CPU). Chapter 7 examines input/output configurations, and chapter 8 explores peripherals.

The *central processing unit* (CPU) comprises three elements: the power supply, the microprocessor (or processor), and memory. We examine each of these elements in turn.

Power Supplies

The power supply is one of the most important elements of the CPU, because without it, processor and memory circuits could not function. It is the simplest element of the CPU to select. Unfortunately, it is also one of the most overlooked elements.

The power supply, in addition to providing power to the processor and memory, often provides power to input/output (I/O) modules and for communication between the processor and remote I/O modules. Because of these requirements, it is often necessary to select a power supply capable of delivering more power than required for operating just the processor and memory.

In most cases, programmable controller (PC) power supplies require ac input voltages. Some, however, use dc input voltages. These are more commonly used in applications in which ac voltage is not available, such as at on-site field operations. Generally, ac-input power supplies accept 120 or 220 volts ac input, while dc power supplies accept 125 or 24 volts dc input.

A block diagram of a typical PC power supply is shown in Figure 6.1. The ac line voltage is stepped down by the transformer and rectified (or converted to dc voltage) in the rectifier module. After rectification, the voltage is filtered and regulated.

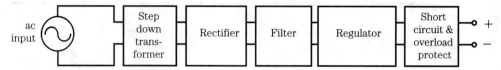

Figure 6.1 Block diagram of a PC power supply.

The *regulator* is an electronic device that maintains the output voltage at a constant level, regardless of the load on the power supply. After the regulator is the short circuit and overload protection circuitry, which prevents a short circuit or an overload from destroying the power supply (or the microprocessor or memory circuits).

Most of the elements found in the power supply shown in Figure 6.1 are found in normal power supplies. Certain differences exist, however, between a normal power supply and a PC power supply. PC power supplies are used in industrial plants where ac line voltage experiences fluctuations, both in voltage level and frequency. These fluctuations are caused by the starting up or shutting down of large pieces of equipment, such as motors and pumps. A normal power supply would not function properly under such fluctuations in voltage and frequency; however, a PC power supply is designed to operate under these conditions. Additionally, PC power supplies are designed to shut the microprocessor and memory down if voltage variations or frequency variations become excessive.

One additional difference between a normal power supply and a PC power supply is that the PC power supply is designed to operate in areas of high *electromagnetic interference* (EMI). Electromagnetic interference is often generated by heavy equipment operating near the PC. PC power supplies are protected from EMI by isolation transformers installed between the ac power line and the programmable controller.

As with standard power supplies, the PC power supply should be sized according to the load that will be connected to the power supply. Many programmable controllers include the power supply as an integral part of the unit. With these PCs, you do not need to select a power supply. For PCs in which the power supply is not included, it is important to select a power supply that can handle the maximum load of the microprocessor and memory, plus part of the load from the I/O modules. For PCs that use many I/O modules, *auxiliary*, or remote power supplies are needed. These supplies should be selected in accordance with the load requirements of the I/O modules.

The easiest way to determine the proper size of the power supply is through *current summation*. You can simply sum the current requirements for all I/O modules to be used and select a power supply capable of supplying the total current, or slightly more than the total current, for the PC and its I/O modules.

PC power supplies often include *battery backup systems*. Battery backup systems are particularly useful for certain types of memories, called *volatile memories*, that do not maintain their programs when the power supply is interrupted. Battery backups employ many different types of batteries, including carbon-zinc, alkaline, nickel-cadmium, lithium, and lead acid. Generally, carbon-zinc or alkaline batteries are used in the majority of PCs, although lithium batteries are becoming very popular. Carbon-zinc, alkaline, and lithium batteries are called *primary* batteries because they are not rechargeable. The nickel-cadmium and lead-acid batteries are called *secondary* batteries because they can be recharged. Using batteries in PC systems requires proper maintenance and monitoring. Batteries that become discharged often leak electrolyte, which can corrode the contacts and wiring in the power supply.

The Microprocessor

At the heart of the microprocessor (or processor) of the PC is the integrated-circuit microprocessor chip. The microprocessor chips used in PCs are exactly the same as those used in computers: generally the Z80, 8080, 8086, 6800, 9900, 286, 386, or 486 family. If you understand how a microprocessor functions in a personal computer, you also understand how a microprocessor functions in a programmable controller. (The arithmetical and logic operations of microprocessors were demonstrated in chapter 4.) The differences between the PC and the personal computer are in the power supply, noise protection circuitry, programming language, and nature of the I/O circuitry.

The microprocessor forms the intelligence of the programmable controller. It is, as previously noted, an integrated circuit that performs mathematical operations, handles data, and executes diagnostic routines required of the programmable controllers. In short, the microprocessor controls all activities of the PC.

The microprocessor is controlled by a system of programs known as the *executive* program. Executive programs are stored permanently in the PC and supervise the operation of the PC, including control, data processing, and communication with I/O modules and peripherals.

In addition to being controlled by the executive program, the microprocessor is also controlled by *application programs*, or *user programs*. These programs are the actual instructions for a particular control application. These programs are usually in the form of ladder-logic diagrams, but can also be written in a higher-level language, such as the languages described in chapter 9.

Some programmable controllers use an approach known as *multiprocessing*, which is simply a method of reducing the time required to implement a program. With multiprocessing, several microprocessors are used to perform control tasks. An example of multiprocessing is the use of an *intelligent I/O module*, such as the *proportional-integral-derivative* (PID) control module. This module is programmed with the equations for PID control and is generally located separately from the CPU. The primary advantage of multiprocessing is a great reduction in the time required to process instructions and control functions.

Microprocessors generally are categorized by word size. As discussed in chapter 4, a word is a group of one or more bytes. Typically, a byte is a group of eight bits; however, words can comprise 4, 8, or 16 bits. Word length is an important characteristic of PCs: a longer word length allows faster manipulation of data because more data are handled in one operation with a longer word.

The microprocessor is responsible for accepting inputs from the I/O modules, manipulating the data from the inputs, and updating outputs. This process is known as *performing a scan*. PCs are also categorized by *scan time* (the length of time required to perform a scan). Typically, scan time ranges between 1 and 100 milliseconds (ms).

Scan time of a PC must be considered when selecting the proper PC for a particular operation. If input signals change more quickly than the scan time, the PC cannot act upon the input data. Thus, it is important that a PC be selected with a scan time faster than changes in input data.

Microprocessors also include *diagnostic programs*. These programs detect failures in communications, system operation, etc. A failure in any one of these areas would be detected by the diagnostics in the microprocessor, activating an alarm circuit that would signal a failure.

Memory

Programmable controllers, by definition, use a programmable memory for storing instructions used to implement specific functions. Memory is thus an important aspect of the central processing unit. This section discusses the types of memory available for use with PCs and the *architecture* of memories used in PCs.

The similarity between PCs and personal computers continues for memory. The types of memories used in PCs are the same as those used in computers.

Memory systems

Memory systems fall into one of two categories: *volatile* or *nonvolatile*. Volatile memory is memory that loses its contents when the power supply is interrupted. Volatile memories are easily altered or erased. Nonvolatile memories retain their contents even if the power supply is interrupted and are usually difficult, if not impossible, to alter. It is not unusual to find both volatile and nonvolatile memories in programmable controllers.

Random-access memory (RAM) is a volatile semiconductor memory. RAM is often categorized by the semiconductor technology used in its construction, such as CMOS RAM for complementary-metal-oxide semiconductor random-access memory. RAM, being a volatile memory, loses its contents when the power supply is interrupted. Thus, the information stored in RAM is easily changed. In other words, RAM memory is easy to reprogram. While random-access memory is not used to store the executive programs, it does work well for storing input data and application programs. RAM is also a relatively fast memory.

Two types of random-access memory exist that are nonvolatile. These are *core memory* and *NOVRAM*, or NOn Volatile RAM. Core memory, shown in Figure 6.2, consists of small magnetic donuts magnetized in either a clockwise or counterclockwise direction, depending on the flow of current in the wires that pass through the center of the donuts. Each donut represents one bit of memory. Magnetization in the clockwise direction represents a 1, while magnetization in the counterclockwise direction represents a 0. Because the donuts can maintain their magnetization for long periods of time, core memory is nonvolatile. Core memory is an old technology, however, and is slower, more expensive, and less compact than modern semiconductor RAM.

NOVRAM employs conventional semiconductor RAM and a nonvolatile semiconductor memory known as *EEPROM*, which is discussed further in later chapters. Both memories are fabricated on a single chip. NOVRAM is not used much in programmable controllers today, and when it is used, it is generally in smaller PCs.

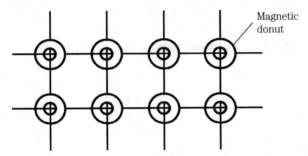

Figure 6.2 Core memory showing two 4-bit words.

All nonvolatile memories (apart from core memory) represent some form of *read-only memory* (ROM). Generally, read-only memory is not easily reprogrammed. Once a program is entered into ROM, it stays there permanently, regardless of the status of power to the ROM chip. Several types of read-only memory exist: PROM, EPROM, UV-EPROM, EEPROM, and EAROM.

PROM (programmable read-only memory) cannot be altered once it has been programmed (once data or instructions have been stored in the memory). To change the contents of PROM memory, the entire PROM chip must be replaced with one that contains the desired data. *EPROM* (erasable programmable read-only memory) chips can be erased in one of two ways: exposure to ultraviolet light, in which case it is called a *UV-EPROM*; or with an electric charge, in which case it is called an electrically erasable programmable read-only memory, or EEPROM.

The erasable and reprogrammable nature of the two EPROMs makes them popular memories. Reprogramming an EPROM is not as easy as reprogramming a RAM, however, because the EPROM must be totally erased before reprogramming can occur. EPROMs also have finite lifetimes; they can be erased and reprogrammed only a limited number of times. Nevertheless, their use in PC memories has increased.

A third type of ROM is the EAROM, or electrically alterable read-only memory. Applying a relatively low voltage to a pin on an EAROM chip erases it. Thus, EAROMs can be quickly erased without being removed from the circuit board. EAROM is used in very few programmable controllers today, however, because it is a relatively slow memory.

Memory architecture

The architecture of a PC's memory specifies how the various memory systems described above are organized and used by the PC to perform control functions. Memory architecture is usually shown schematically by a *memory utilization map*, such as the one shown in Figure 6.3. The memory map in Figure 6.3 indicates that PC memory is divided into three major types: system memory, input/output status memory, and application memory.

System memory can be subdivided into two types of memory: *executive memory* and *scratch-pad memory*. Executive memory contains the executive program (or executive operating system). Because the executive program is provided by the PC

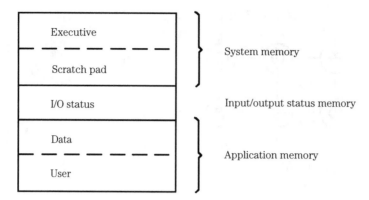

Figure 6.3 Memory utilization map.

manufacturer and rarely changes, executive memory is ROM (or, more specifically, PROM). It was stated earlier that the executive program controls the operation of the PC. It provides the translation between the high-level programming language (such as BASIC) and the binary machine language, scans the PC to update system status, and reads inputs and updates outputs. In administering these functions, the executive often needs an area of memory to store data temporarily. The portion of memory reserved for this purpose is called the scratch-pad memory. Scratch-pad memory consists of RAM and is reserved for the exclusive use of the executive; it is not accessible to the PC user.

Input/output status memory is accurately described by its name. It is a portion of RAM set aside for storage of current I/O statuses. Because the executive program requires I/O status updates, the I/O status memory can be thought of as a part of the system memory.

Application memory is the final area of PC memory. Like system memory, application memory can be subdivided into two types: *data memory* and *user memory*. These memories hold the data used by the microprocessor to fulfill its control functions along with the user program (or instructions) that direct the microprocessor to perform its control functions.

Data memory generally holds preset values. For example, the user program could contain a statement that sets a counter or timer to a particular value, based on I/O status. The program could also instruct the microprocessor to perform certain math functions (addition, subtraction, etc.) on data obtained. The data memory would then store the preset values for timers and counters and the instructions for the data manipulation (math functions). Because the values stored in data memory must be changed frequently, data memory is RAM.

User memory is the memory most accessible to the user or programmer. The user program provides specific instructions for control of processes and provides the "programmable" feature of programmable controllers. User memory is scanned by the microprocessor for specific instructions when the microprocessor is directed to scan by the executive. This scan of, or communication with, user memory is accomplished through the use of two *buses*. A bus is a group of lines used for data transmission or power distribution. The first bus addresses a particular location in memory and is known as the *address bus*. The second bus is used to transmit the data stored in memory and is called the *data bus*. Because the user program might change frequently, user memory is RAM.

Although RAM is the predominant memory type for user memory, EPROM and EAROM are also sometimes used. Both EPROM and EAROM are not as easily reprogrammed as RAM; therefore, EPROM and EAROM are used in PCs whose user programs seldom, if ever, change.

Not all PC manufacturers use the same type of memory utilization map shown in Figure 6.3. Two alternate examples are shown in Figures 6.4 and 6.5. These are designated as memory maps for code addressing and external data addressing, respectively, and correspond to system memory and application memory. Note the use of hexadecimal codes (left) and decimal codes (right) in the figures. You can determine the nature of the memory map if you read the literature supplied with the PC carefully.

General comments on memory

The basic structural units of memory—the bit, byte, and word—were discussed previously. While microprocessors are generally characterized by word size, memory is characterized by the number of words that can be stored. The amount of memory is designated by the letter K. One K of memory provides the capacity for the storage of 1024 words. Likewise, 8K of memory provides the capacity for the storage of 8192 words. The words can vary in length and are typically either 8-bit or 16-bit. Word length of course affects the total memory capacity of the PC. For example, a 1K memory using 8-bit word length can store only one half the information stored in a 1K memory using 16-bit word length.

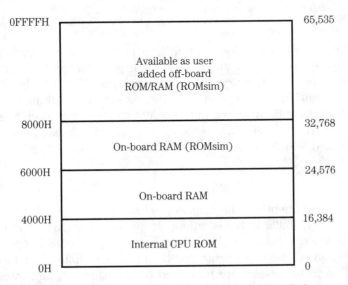

Figure 6.4 Memory map for code addressing (reprinted with the permission of Blue Earth Research, © 1990,1991).

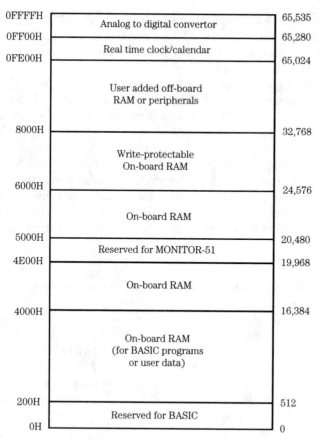

0FFFFH	Analog to digital convertor	65,535
0FF00H		65,280
0FE00H	Real time clock/calendar	65,024
	User added off-board RAM or peripherals	
8000H		32,768
	Write-protectable On-board RAM	
6000H		24,576
	On-board RAM	
5000H		20,480
4E00H	Reserved for MONITOR-51	19,968
	On-board RAM	
4000H		16,384
	On-board RAM (for BASIC programs or user data)	
200H		512
0H	Reserved for BASIC	0

Figure 6.5 Memory map for external data processing (reprinted with the permission of Blue Earth Research, © 1990,1991).

Memory is an important consideration in the specification of a microprocessor. For example, the scan time of a microprocessor is generally specified in terms of time per K of programmed memory. When memory is specified for a PC, it is important that the distribution of the memory be understood. Although many PC suppliers specify only application memory, others might include system memory and I/O status memory in the total amount of memory specified. The supplier's literature should be consulted to determine the exact category of memory specified.

Summary

The three basic components of the central processing unit were discussed. These components are very similar to the components found in a personal computer. Certain differences do exist, however. Power supplies for programmable controllers are generally more rugged and more protected than power supplies found in personal computers. PC microprocessors are protected from noise and other interferences. Memory requirements for PCs are somewhat different from memory requirements for personal computers, especially regarding input/output status memory.

Input/Output Interfaces

Assume we want to use a programmable controller (PC) to control steam temperature. A relay control circuit for that purpose was provided in Figure 5.6. As a first attempt, we use the circuit of Figure 5.6 but replace the temperature-activated switch (comparator) and relay with a PC. The result is shown in Figure 7.1. After several hours, we notice that the pump motor never started. We measure the steam temperature in the line and discover that it is dangerously high. We conclude that the PC circuit does not work, and we manually turn on the pump motor.

Why did the PC circuit in Figure 7.1 not perform as expected? Two very good reasons exist. First, the output of a thermocouple is typically on the order of millivolts. The PC in Figure 7.1, however, uses 0 volts dc as a 0 level and +5 volts dc as a 1 level, so the output of the thermocouple is insufficient to provide a 1 input to the PC. Second, the PC is incapable of driving the pump motor because the PC does not provide enough output power, and it cannot be connected directly to an ac source.

For the circuit in Figure 7.1 to perform properly, the input signal to the PC must be boosted to a level compatible with the CPU's logic circuitry, and the output signal from the CPU must be isolated from the ac source and boosted to an appropriate power level. These functions are performed by *input/output interfaces*, as shown in Figure 7.2. The I/O interfaces thus provide the means by which the PC interacts with input and output devices.

This chapter explores the various I/O interfaces used with PCs. Generally, these interfaces fall into one of the following classes: discrete, numerical data, analog, or special.

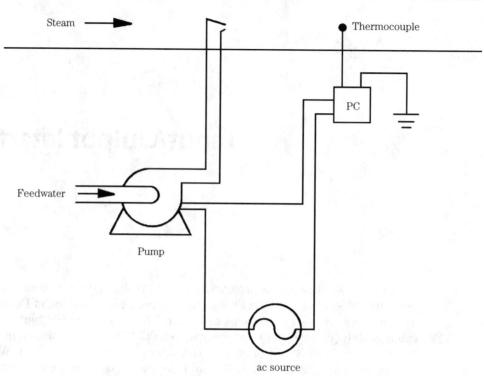

Figure 7.1 Figure 5-6 redrawn using a PC.

Discrete Interfaces

The *discrete interface* is the oldest and most commonly used class of I/O interface. It is used with input/output devices that provide or require discrete signals. For example, a discrete input device is a switch that is either on or off. Similarly, a discrete output interface connects the PC to an output device that can be in only one of two states (on or off), such as lights, relays, or alarms.

One of the primary functions of an input interface is to match voltage levels between an input device and the PC. A block diagram of an input interface that performs this function is shown in Figure 7.3. The interface in this figure accepts an ac input voltage signal and converts it to a dc level suitable for the PC's microprocessor. For example, the input might be from a device that provides a 120-volt ac signal. The first stage of the interface rectifies and filters the signal. The second stage of the interface is the signal threshold detector. If the ac signal is greater than the background noise level (i.e., if the signal is sufficient to indicate that the input sensor is functioning in the on mode), the signal is treated as a 1 signal, and an on voltage at a level appropriate for the PC's microprocessor is generated. The final stage of the interface is the isolation stage, which could be a transformer or optical coupler. The isolation stage removes any electrical connection between the input device and the PC, preventing any damage to the PC that could occur from anomalies such as voltage spikes.

A discrete output interface is shown in Figure 7.4. As the figure shows, the output interface is similar to the input interface. The signal from the PC is electrically isolated from the output device. The signal enters a switching section that converts the PC's dc signal to an ac voltage at a level suitable for use by the output device.

Although Figures 7.3 and 7.4 show input/output interfaces that convert ac to dc and dc to ac, respectively, not all discrete I/O interfaces are used to convert between ac and dc. In many cases, the input and output devices connected to the PC use solely dc voltage; however, the voltage level might not be the same as that required for the operation of the PC. Additionally, current inputs and outputs might be required. In these instances, the I/O interfaces would be used to convert the dc voltages to appropriate levels, or to convert current inputs to voltages for use by the PC and voltage outputs from the PC to currents for use by the output devices. Other I/O interfaces are used to handle transistor-transistor logic (TTL) voltages (5 volts dc).

Numerical Data Interfaces

Discrete interfaces handle I/O data that consist of single bits: either 1 or 0. In contrast, *numerical data interfaces* handle inputs and outputs that consist of many bits. In other words, numerical data interfaces are capable of handling data consisting of multiple bits, such as binary-coded decimal (BCD) inputs and outputs.

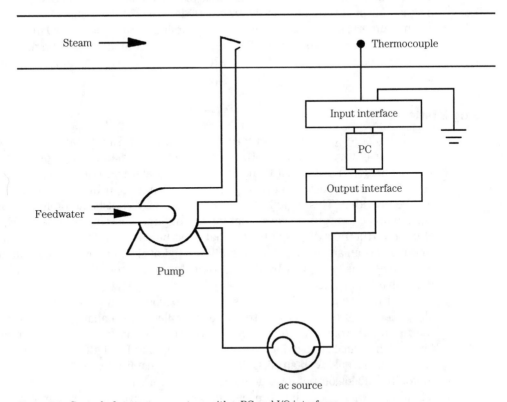

Figure 7.2 Control of steam temperature with a PC and I/O interfaces.

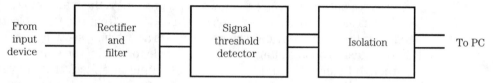

Figure 7.3 Block diagram of a discrete input interface.

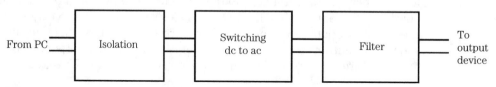

Figure 7.4 Block diagram of a discrete output interface.

A typical multibit input would be data from a thumbwheel switch. A BCD thumbwheel switch set to the number 9, for example, would generate the BCD number 1001. These four bits would be processed by the input interface and transferred to the PC in a format acceptable to the PC. An example of a multibit output device would be a seven-segment display, which requires four bits of data to light the display.

The numerical data I/O interface is very similar to the discrete I/O interface in that it handles discrete bits of data. The difference, of course, is that the numerical data I/O interface handles multiple bits, whereas the discrete I/O interface handles only single bits.

Analog Interfaces

Many input devices, such as thermocouples and pressure transducers, provide a continuously varying analog output signal. For these signals to be processed by the PC, the analog signals must be converted to digital. In addition, once the PC has responded to the information received from the input sensor, a digital output signal will be provided. This digital PC output signal might not be suitable for controlling certain analog output devices, such as motors and chart recorders. Thus, in these situations, the PC's digital output signal must be converted to analog. I/O interfaces that convert an analog signal to digital or a digital signal to analog are called *analog interfaces*. The two types are digital-to-analog converters (DAC) and analog-to-digital converters (ADC).

A typical analog input interface is the thermocouple interface. Thermocouples usually generate output signals on the order of millivolts. An analog input interface for a thermocouple would thus be capable of accepting a very low-level analog input signal, filtering the signal, amplifying it to a level appropriate for the PC, and then converting it to a digital signal with an ADC. If the output signal from the PC controls an electric motor, the analog output interface of the PC converts the output signal to analog (with a DAC) and then boosts it to the level necessary to drive the electric motor.

As these examples point out, analog I/O interfaces are essentially DACs or ADCs that include amplification and filtering capabilities.

Special Interfaces

The discrete, numerical data and analog interfaces satisfy the majority of data processing requirements for input and output signals of the PC. There are a few special interfaces, however, that deserve mention. These interfaces are sometimes called *intelligent interfaces* because they perform certain tasks independent of the PC. The use of intelligent interfaces in process control is often known as *distributed processing*. A few of these special interfaces are discussed in this section.

Data-processing interfaces are modules that perform microprocessor functions. A data-processing interface can be thought of as a small, dedicated computer that performs preset calculations on data received from sensors or other devices connected to the interface. The interface would be capable of storing data and would perform mathematical calculations (such as addition or subtraction of input signals, square roots of input signals, etc.) on the data. The use of data processing interfaces eliminates those demands from the PC's microprocessor, thus freeing the PC to handle larger control functions.

As seen in the next chapter, peripherals constitute an important part of the PC system. The use of peripherals, such as printers and video display terminals, requires special interfaces. A typical interface for peripherals would be an ASCII interface, which transmits alphanumeric data between the PC and its peripherals. The ASCII interface is not an intelligent interface, however, because it does not perform microprocessor or control functions. It does, however, speed up communications between the peripherals and PC.

One of the most popular of the special I/O interfaces is the *proportional-integral-derivative* (PID) interface. The PID interface provides PID control for various processes that require continuous, closed-loop feedback control. PID control is a very important concept in PCs and is discussed in detail in the following section.

Proportional-integral-derivative control

PID control can best be described with an example. Figure 7.5 shows a system in which temperature must be controlled. A reactant stream is fed to a reactor, which converts the reactant into a product. If the temperature of the reactant stream is less than a specified temperature, T, the yield of product produced by the reactor is diminished, and the reactor becomes uneconomical to operate. At a temperature greater than T, the reaction becomes vigorous, eventually approaching an explosive level. Thus, economic and safety considerations require that the reactant stream be maintained at or near temperature T.

The simplest method of controlling reactant stream temperature T is shown in Figure 7.6. In this figure, a set-point temperature T_s is fed to a proportional controller, as is the actual temperature T of the reactant stream. (The proportional controller can be an intelligent interface or a PC programmed for proportional control.) The

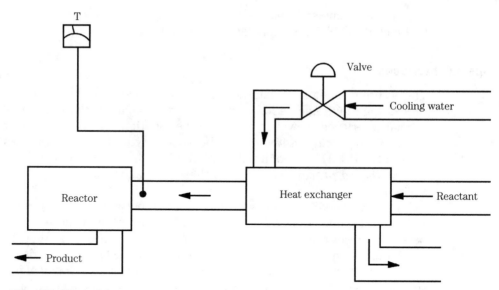

Figure 7.5 The model system.

proportional controller measures the error ε in the temperature, where $\varepsilon = T_s - T$. The output of the proportional controller is given by equation 7-1:

$$e = K\varepsilon + e_c \qquad\qquad\qquad (7\text{-}1)$$

where

e = voltage output from the proportional controller (the control signal)
K = gain
ε = error signal = $T_s - T$
e_c = constant
T_s = set-point temperature (desired temperature)
T = measured temperature

From equation 7-1 you can see that the control signal e is proportional to the error signal ε. In fact, equation 7-1 is just the equation for a straight line. Thus, the control signal is linearly proportional to the error signal.

The temperature-versus-time plot for the proportionally controlled system in Figure 7.6 is shown in Figure 7.7, and the plot is compared to that of an uncontrolled system. It is assumed that the proportional control system is *ideal*; i.e., equation 7-1 is obeyed exactly. It can be seen from the figure that, with no control, the temperature of the reactant stream increases constantly, eventually approaching the explosive limit. With proportional control, however, the temperature oscillates back and forth about the set-point temperature, eventually reaching equilibrium at T_s.

The oscillatory approach to the set-point temperature is caused by time lags in the system. The control valve that allows cooling water to flow to the heat exchanger does not open instantaneously, and the heat exchanger does not drop the reactant stream temperature immediately. Because of the time lags in the valve and heat exchanger operations, the controller overcorrects for temperature, driving the reactant stream temperature below the set-point value. The controller then senses the new error and attempts to compensate for it. Again, it overshoots the mark, but not by quite as much as before. This process continues until the reactant stream temperature is very nearly equal to the set-point temperature. The action of the controller as shown in the figure thus appears to be that of a damped oscillator.

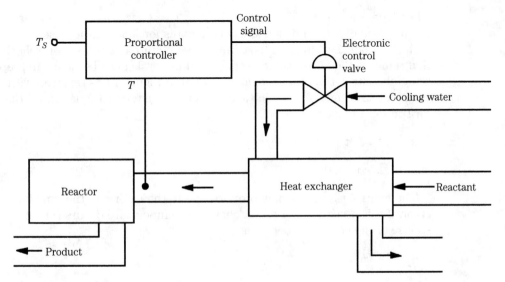

Figure 7.6 The proportional control system.

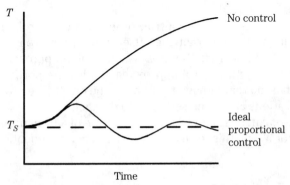

Figure 7.7 Comparison of ideal proportional control with no control.

Figure 7.8 shows the response of a real (i.e., nonideal) proportional controller. As can be seen, the temperature of the reactant stream reaches a steady state value higher than that of the set-point temperature. The difference between the steady state and set-point temperatures is called the *offset*.

The offset occurs because equation 7-1 is not followed exactly. Equation 7-1 is the steady-state control equation. It assumes that the change in temperature from the setpoint value occurs instantaneously, which is not usually the case. The effect of the finite time required to change the temperature is the addition of a term to equation 7-1, as shown in equation 7-2:

$$e = K\varepsilon + e_c + f(t,K) \tag{7-2}$$

where the function $f(t,K)$ is called the *transient response*. The transient response is a function of time and gain and is largely responsible for the offset.

In the model system, an offset cannot be tolerated. A better control scheme, one that removes the offset, is the proportional-integral control scheme. In proportional-integral control, the proportional controller in Figure 7.6 is replaced with a proportional-integral controller, which adds an additional term to equation 7-1 that corrects for transient response:

$$e = K\varepsilon + K/\tau \int_{t=0}^{t} \varepsilon \, dt + e_c \tag{7-3}$$

where τ is a time delay.

The control signal e is now proportional to the error and the time integral of the error. If it is assumed that a unit change in temperature occurs (i.e., $\varepsilon = 1$), the response of the controller becomes

$$e = K + (K/\tau)t + c \tag{7-4}$$

where c is a constant that is determined by initial conditions. Thus, with a unit change in temperature, e changes by an amount equal to K (proportional action) and then changes linearly with time at a rate equal to K/τ. In this way, the offset resulting from transient response is removed.

Figure 7.9 shows a comparison of no control, proportional control, and proportional-integral control. It can be seen that the offset in temperature is removed by proportional-integral control and that the steady-state value approaches the set-point value. Proportional-integral control approximates ideal proportional control, with the same degree of oscillatory behavior. Although the offset is removed, it takes a long time for the oscillations about the set point to dampen.

The oscillatory behavior can be removed by the inclusion of a term proportional to the derivative of the error. When this term is included, the control equation becomes the following:

$$e = K\varepsilon + K/\tau \int_{t=0}^{t} \varepsilon \, dt + K\tau \, d\varepsilon/dt \tag{7-5}$$

Control signal e is now proportional to the error signal, the integral of the error signal, and the derivative of the error signal. The first term in equation 7-5 corrects for

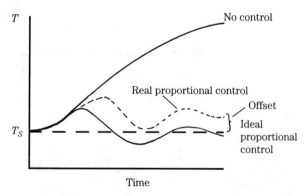

Figure 7.8 Comparison of real proportional control with ideal proportional control and no control.

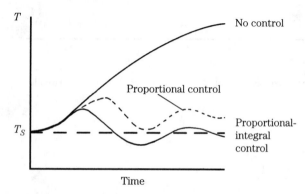

Figure 7.9 Comparison of no control, proportional control, and proportional-integral control.

the difference between the actual and set-point temperatures, the second term eliminates the offset, and the third term anticipates the change in T and dampens out the oscillations. This equation defines the PID control scheme. A comparison of no control, proportional control, proportional-integral control, proportional-derivative control, and proportional-integral-derivative control is shown in Figure 7.10.

A PID interface implements equation 7-5 to control various processes. PID control can also be performed by a PC programmed with the PID control algorithm; however, using a PID interface relieves the PC of one of its functions, allowing the PC to control other process steps.

Summary

Without input/output interfaces, the PC would be unable to perform its control functions. The basic types of interfaces were described, including discrete, numerical data, analog, and PID interfaces.

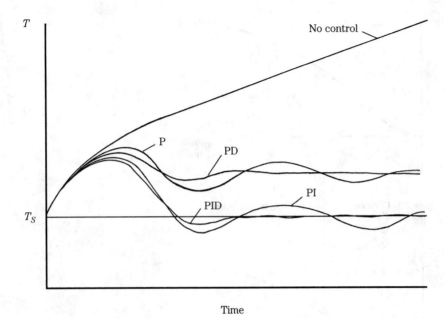

Figure 7.10 Comparison of the proportional-integral-derivative control scheme with all others.

Peripherals

Throughout this book comparisons have been drawn between programmable controllers (PCs) and personal computers. Such comparisons are particularly useful in highlighting the functions of *peripherals*, or the external devices connected to the PC. A very basic personal computer system consists of four components: the computer itself, or central processing unit (CPU); a keyboard for data and program entry; a video-display terminal (VDT) to display the program or system status; and a line printer for hard copies of programs or data. In this simple system, the keyboard, VDT, and printer are all personal computer peripherals.

PCs also use peripherals that perform functions similar to those for personal computers. There are three basic classes of PC peripherals: programming and data entry devices, display devices, and documentation devices. This chapter discusses each class of device. Additionally, the important topic of communications between peripherals and the PC is also discussed.

Programming and Data Entry Devices

As the name implies, programming and data entry devices allow the PC operator to enter programs, preset values, and such into the PC's memory. Many of these devices are on the market today. *Cathode ray tubes* (CRTs) can be thought of as a combination of the video display terminal and the keyboard of the personal computer system. The CRT contains all the electronic circuitry necessary for communication with the PC. CRTs are extremely useful: the display can present large amounts of information (limited only by the size of the screen), and the presence of a complete keyboard allows easy access to the PC's memory. While some CRTs are stationary, others are portable.

Two basic types of CRTs are available: dumb and intelligent. Of the two types, the dumb CRTs are older, less expensive, and more flexible in terms of compatibility with a wide variety of different PCs. Dumb CRTs do not contain their own microprocessor or memory. They must be connected with the PC before they can operate properly. They are used as terminals and allow online programming only. Intelligent CRTs, on the other hand, contain their own microprocessors and memories. An intelligent CRT allows the programmer to write and edit programs without being connected to the PC, which is called *offline programming*. Intelligent CRTs are much more expensive than dumb ones and are usually not compatible with more than one brand of PC. Personal computers can be used as data entry and programming devices; they can thus function essentially as intelligent CRTs.

For some small PCs, *miniprogrammers* offer an inexpensive, portable alternative to the CRT. These are handheld devices that resemble pocket calculators, except that their displays are somewhat larger, and the keyboards contain special programming keys. Miniprogrammers can also be either dumb or intelligent. They are used to enter programs into the PC's memory and to edit programs. Generally, miniprogrammers are compatible only within a manufacturer's family of PCs.

Program loaders are devices used to load or reload a program into the PC's memory. Program loaders are not programming devices; they are not used for writing or editing programs. They merely provide a convenient method for loading a program into the PC's memory. One of the oldest and most popular program loader is the cassette tape recorder. A written and edited program is fed into the cassette tape recorder, which is then transported to the PC site, where its contents are unloaded. Another program loader gaining in popularity is the memory module, which consists simply of an EPROM (or EEPROM) and the associated electronic circuitry. Written and edited programs are read into the EPROM memory, which is then transported to the PC site where its contents are unloaded into the PC. Although program loaders have always been popular with small PCs, they are now gaining popularity with larger PCs.

Some small, relatively inexpensive PCs use ROM, PROM, or EPROM chips for memory. For these PCs, a *memory burner* is used to impress a program onto the memory chips. These memories, as discussed in chapter 6, are not as easily reprogrammed as RAM. Thus, memory burners do not find much use in PCs today.

Thumbwheel switches are convenient devices for inputting certain data into the PC. A thumbwheel switch is a 10-position switch, each position of which corresponds to a number from 0 to 9. When a thumbwheel switch is tuned to a particular number (such as 7), it transmits the binary-coded decimal (BCD) code for that number (0111 for the decimal number 7, for example). Thumbwheel switches are most often used to input preset values (for timers and counters) and high and low limits. The thumbwheel switch is connected to the PC by four wires, one for each BCD bit.

Display Devices

In addition to serving as a programming and data entry device, the CRT is an excellent display device. Because of its large screen, many program steps and data points can be displayed simultaneously. Additionally, CRTs can be programmed to display process steps diagrammatically, i.e., process status can be displayed as a flow diagram. Using color terminals also greatly enhances the utility of the CRT.

Two other display devices are often used. The *seven-segment display* is used to display numbers that correspond to speeds, times, or distances, depending on the process. This display uses seven-segment light-emitting diodes (LEDs) or liquid-crystal displays (LCDs). Some PC manufacturers are providing *intelligent-alphanumeric displays*, which are programmable displays that, in addition to displaying numbers, can display messages or warnings programmed into its memory.

Documentation Devices

The *line printer* is the principal documentation device for PCs. Some manufacturers offer special documentation systems, such as a system that translates ladder diagrams into written programs, and vice versa, but the line printer is still an essential element of these systems.

Three printer characteristics are of importance: baud rate, buffer size, and printing speed. *Baud rate* defines the number of bits per second of binary data that can be received or transmitted during serial communication between the PC and a peripheral. (The serial transmission of data is discussed in the next section.) Baud rates vary widely. They can be as low as 50 for special-purpose, low-speed PCs, or as high as 19,200 baud. The most common rates are 300, 1200, and 9600 baud. When selecting a line printer for use with a PC, care should be taken to ensure that the printer can work at the PC's particular baud rate. Usually, printer baud rates are adjustable over a small range.

A printer's *buffer* is actually a short-term memory. If the PC transmits data faster than the printer can print the data, the data not yet printed are stored temporarily in the printer's buffer. Buffer size should thus be considered along with the printing speed of the printer, which is given in characters printed per second (cps). Here again, printing speeds vary widely. A typical slow speed (daisy wheel) printer prints 20 characters per second, while printing speeds of 200 characters per second can be achieved using dot matrix printers.

Communications

Two general schemes are used by which bits of data are transmitted from the peripheral to the PC and vice versa: *parallel transmission* and *serial transmission*. These two schemes are discussed in the following subsections. Serial transmission is most often used in PC systems, but parallel transmission is discussed for completeness.

Parallel transmission

The principles behind parallel transmission are demonstrated in Figure 8.1. Data, in the form of binary bits, are stored in the PC logic gates. These bits are not transmitted until a transmit pulse is received by the gates. When the pulse is received, the bits are transmitted via the transmission lines to the gates in the peripheral device. Note that the transmit pulse is also sent to the peripheral. Upon receipt, the gates are reset, which means they are cleared of previous data and readied to receive the transmission.

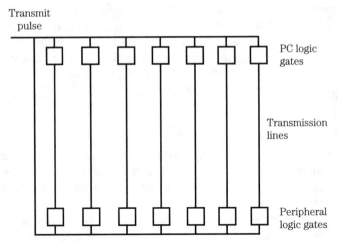

Transmit pulse

PC logic gates

Transmission lines

Peripheral logic gates

Figure 8.1 Parallel transmission.

The most interesting feature of Figure 8.1 is the number of transmission lines involved. In parallel transmission, each bit from each gate in the PC is carried by a separate line. For a large PC, this means that several lines are needed for communications between the PC and each peripheral. One bit per line is one of the disadvantages of parallel transmission and effectively restricts the use of parallel transmission to short distances of 6 feet or less. Another disadvantage of parallel transmission is that, because each line is dedicated, parallel interfaces are not usually interchangeable with interfaces for different peripherals.

The primary advantage of parallel transmission is its speed. Because many bits of data are transmitted simultaneously along several transmission lines, parallel transmission is unmatched for its speed in communicating data. Unfortunately, the speed of transmission is often faster than the speed at which the peripheral can process the data. For example, a printer that prints only 20 characters per second does not require the rapid transmission of data.

Note that in Figure 8.1 a transmit pulse is used to both transmit data from the PC to the peripheral and reset the peripheral's logic gates. It is essential to synchronize this transmit pulse. Synchronization is obtained using a clock in the PC, which provides transmit pulses at regular intervals. Note also that the entire transmission from the PC to the peripheral occurs during one clock pulse. That is, the transmit pulse starts the transmission, and the transmission is completed during the time interval between transmit pulses.

Serial transmission

The principles behind serial transmission are demonstrated in Figure 8.2. Data from the PC logic gates are accepted one at a time by the driver. Pulses from the master timer prompt the acceptance of data bits from the PC. When each bit of data is received by the driver, it is driven, or transmitted, along a single transmission line to the receiver. The receiver directs each bit to the correct peripheral logic gate, in step with a pulse from the timer. Notice that a line connects the master timer with the timer. This line synchronizes the two timers so that each received bit is directed to the correct gate. The driver and receiver shown in Figure 8.2 are combinations of AND and OR gates activated sequentially by timing pulses from the master timer or the receiver timer.

The most interesting feature of serial transmission is that only one transmission line is used. All bits of data are transmitted one at a time over the one line, rather than all at once over several lines, as in parallel transmission. Thus, serial transmission is slower than parallel transmission. The advantage of serial transmission, however, is that because only one line is used, it allows data to be transmitted over longer distances than achievable with parallel transmission.

Serial transmission is used most often in PC systems because the peripherals used in PC systems are generally slow (sometimes as slow as 110 baud) and require long cable connections. A typical serial interface allows data to be transmitted over a distance of 50 feet, compared to 6 feet for parallel transmission.

In most PC systems, data are transmitted in the form of ASCII code. Figure 8.3 shows the pulse train for the ASCII code for the comma character (binary 0101100), which is compared to the pulse train from the master timer. The first pulse from the master timer signals the beginning of the transmission. The binary digits 0 and 1 are

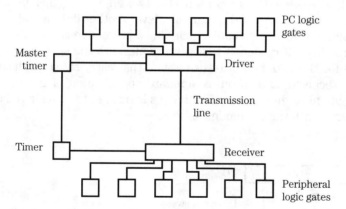

Figure 8.2 Serial transmission.

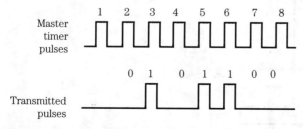

Figure 8.3 Serial transmission of binary number 0101100 (ASCII comma).

represented by zero and positive voltage levels, respectively, and are transmitted in step with the pulses from the timer. A peripheral, such as a printer, that receives the ASCII code 0101100 recognizes the signal as a comma. A seven-digit ASCII code exists for each number, letter, and symbol on the PC's keyboard.

The system shown in Figure 8.2 allows transmission of data from the PC to the peripheral only. This is called *unidirectional*, or *simplex*, transmission. If the driver at the PC and the receiver at the peripheral are both replaced with combination driver/receivers (as shown in Figure 8.4), the form of transmission is *half-duplex*. Half-duplex transmission allows data to be transmitted from the PC to the peripheral and from the peripheral to the PC; however, data transmission occurs in only one direction at a time. To acquire the ability to transmit data simultaneously in two directions, the *full-duplex* system shown in Figure 8.5 is needed. Note that the full-duplex system requires two lines for the simultaneous two-way transfer of data.

Serial interfaces

In the preceding discussion, serial transmission was introduced as requiring only one transmission line. A glance at Figure 8.5, which describes full-duplex transmission, indicates that this explanation was a simplification. In addition to the transmit and receive lines, a line is needed to synchronize the two timers. A signal ground is also needed (but not shown in any of the figures). Several other lines are also typically contained in a serial transmission cable, including a protective ground line and control lines. Control lines carry signals that send requests to transmit data, that carry all-clear signals for the initiation of transmission, that carry signals indicating that data have been received, and so on. A standard specification of the number and physical arrangement of these transmission lines is needed to provide full compatibility of peripherals with various models of PCs.

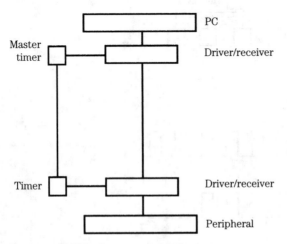

Figure 8.4 Half-duplex transmission.

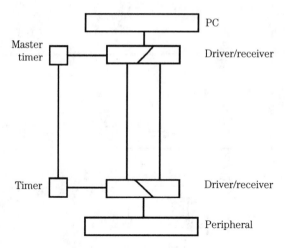

Figure 8.5 Full-duplex transmission.

Several *communication standards* specify signal levels and physical details of the interface, or transmission cable. The four that find most common usage in PC systems are the RS-232C, the RS-449, the 20-mA current loop, and the IEEE 488. The RS-232C and RS-449 are officially proclaimed as standards by the Electronic Industries Association (EIA). The IEEE 488 is officially proclaimed as a standard by the Institute of Electrical and Electronics Engineers (IEEE). The 20-mA current loop is not officially defined by any organization, but has become a standard through common usage. Of these four, the most popular in PC systems is the RS-232C.

The physical details of these standard interfaces can be obtained from the EIA or the IEEE, or from standard reference books, and are not examined here. Note that even a standard such as the RS-232C interface can have many variations. For example, the full RS-232C interface contains 25 lines; however, in most instances, peripherals require only three to five of the lines for proper operation.

Multiple peripherals

Serial communication between a PC and several peripherals can be achieved using either the *daisy chain* or *star* configuration. Both are shown in Figure 8.6. These two configurations offer the advantages of reduced material and labor costs for large systems with peripherals at several locations.

Summary

This chapter presented the three basic classes of PC peripherals, which are quite similar to those associated with personal computers. The basics of parallel and serial communication between peripherals and the PC were reviewed. Note that, of the two, serial communication is used most often in PC systems.

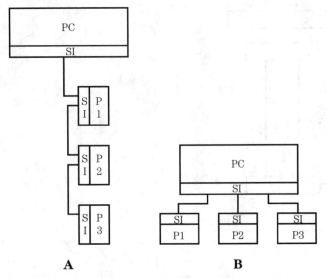

Figure 8.6 PC-peripheral configurations. A) Daisy chain. B) Star. SI stands for *serial interface*. P represents peripherals.

9

PC Software

The software used with a programmable controller (PC) is perhaps the most important part of the PC system for the everyday user. It is the software that allows the user to instruct the PC to perform its functions and that translates the PC's electrical signals into a format the user can understand. Two general classes of languages are used in PC software: *low-level* (or *basic*) languages and *high-level* languages. Both of these are discussed in turn.

Programming languages vary widely in their details, even though they are usually quite similar in their nature. This fact is especially true of the languages employed by different PC manufacturers. It is not the purpose of this chapter to familiarize you with the whole range of programming languages available today; rather, this chapter introduces you to the fundamentals of PC programming languages. Specific examples of programming languages from certain suppliers are used for illustrative purposes only.

Low-Level Languages: Ladder Diagrams

In chapter 5, relay-ladder and contact-ladder diagrams were introduced. The contact-ladder diagram is the most widely used low-level PC language. Review the section of chapter 5 entitled "Ladder diagrams," if necessary.

Figure 9.1 shows the contact symbols introduced in chapter 5. There are three basic symbols: the normally open (N.O.) contact symbol, the normally closed (N.C.) contact symbol, and the coil (or output) symbol. The other symbols shown in Figure 9.1 are simply modifications of these three basic symbols.

The symbols shown in Figure 9.1 are insufficient to program a PC in ladder diagram language; several additional symbols are necessary. These additional symbols fall into one of two categories: basic programming symbols or extended (or enhanced) programming symbols. The basic programming symbols are shown in Figure 9.2. In addition to the N.O., N.C., and output symbols presented previously, five

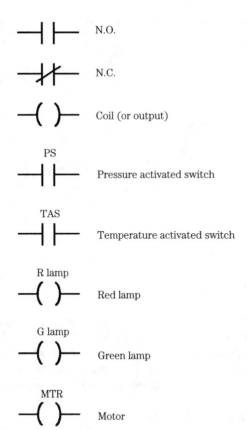

Figure 9.1 Contact ladder symbols used in chapter 5.

other symbols are shown. The first three (out NOT, latch out, and unlatch out) are commonly known as *relay-type instructions* because they were developed for operations with relays. The last two symbols (timer and counter) are instructions that, although basic to the operation of the PC, are not associated with relay operations.

The first three symbols in Figure 9.2 were discussed in chapter 5 in conjunction with relays. Because the PC was designed to replace relays in control applications, these three symbols have slightly different meanings with respect to PC programming. The symbol for the N.O. contact indicates that a specific signal is needed to close the contact and complete the path for the current to flow through the rung of the ladder. The contact symbol does not necessarily represent a contact in PC programming. It might, for example, represent an input signal. If an input signal is present, the contact is closed and current proceeds to flow through the rung. In the absence of an input signal, no current flows through the rung. The inverse holds true for the N.C. contact. The out (or output) symbol is extremely general. It might represent a lamp, motor, coil, or any other device energized when the current path through the rung is completed.

The three remaining relay-type instructions require some explanation, as do the timer and counter instructions. The relationship between the out NOT and the out symbols is similar to the relationship between the N.C. and the N.O. contact symbols. An out NOT symbol indicates that the output is de-energized whenever a current path through the rung is completed. In other words, the output is in the ON condition when no continuity exists through the rung and is in the OFF condition when continuity does exist through the rung. This set of conditions is the inverse of conditions for an out symbol.

The latch out and unlatch out symbols are derived from the *latching relay*. The latching relay is a retentive relay that contains two coils: the latch coil and the unlatch coil. When the latch coil is activated, the N.O. contacts close and the N.C. contacts open. These contacts remain in this condition even if the latch coil is de-energized. When power is supplied to the unlatch coil, the contacts return to their normal positions—the N.O. contacts open and the N.C. contacts close. In other words, when the latch coil of the relay is activated, the coil (for all practical purposes) behaves as if it is energized until power has been supplied to the unlatch coil. The same holds true for the unlatch coil: the relay remains in the unlatched state until power is supplied to the latch coil.

Because the latching relay contains two coils, it requires two rungs on the ladder diagram, as shown in Figure 9.3. In this figure, the outputs of the two rungs represent the same relay and are designated by the same number (001). When contact 2 is closed, the output is latched. The output remains latched until contact 3 is closed. Latching relays are not found in PCs, of course. The latch out and unlatch out symbols merely represent PC output conditions that simulate the action of the actual latching relay.

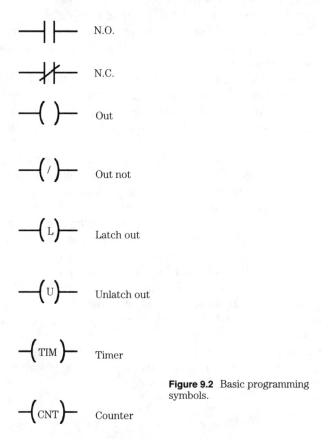

                                              N.O.

                                              N.C.

                                              Out

                                              Out not

                                              Latch out

                                              Unlatch out

                                              Timer

Figure 9.2 Basic programming symbols.

                                              Counter

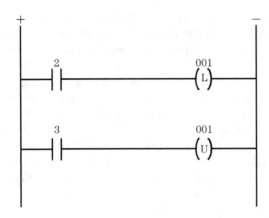

Figure 9.3 Latch and unlatch outputs.

The two remaining basic programming symbols are the timer and the counter. These two instructions are similar; both are used to either energize or de-energize a device or output after a specified time interval or count. When the timer is activated, it begins counting pulses (or time intervals) until the counted time equals a preset value. The counter, on the other hand, keeps count of events, such as pulses or signals. In the example given in chapter 5 of the weighing platform at a shipping station, a counter could be used to keep track of the number of completely filled cartons. The preset value for this counter would be the number of cartons necessary to fill a shipping skid. Counters can count either up or down; that is, they can count up to a preset value or count down from a preset value.

Basic programming symbols vary from manufacturer to manufacturer. For example, some PC suppliers use a circle to denote an output, rather than the parentheses used in this chapter. These differences, however, are usually minor. The user's manual for the PC should be consulted to determine the appropriate programming symbols for a given PC.

The most widely used extended programming symbols are shown in Figure 9.4. The first four symbols indicate *arithmetical operations* (addition, subtraction, multiplication, and division). These four operations can be performed on constants, data stored in memory, or a combination of the two. When an arithmetical operation, such as addition, is performed on two constants, the ladder diagram is similar to that shown in Figure 9.5. The rung in Figure 9.5 represents the addition of the two numbers 23 and 14. The contact on the far left is called the *control contact*. When the control contact is closed, addition is performed on the two numbers.

An arithmetical operation on data stored in memory requires the fifth symbol shown in Figure 9.4, the GET instruction. The GET instruction includes a number above the contact symbol. This number is very similar to a post office box number— it is the location (or address) in the PC's memory where a particular piece of data is stored. For example, a temperature sensor might be directed to store its output in memory address 101. Later, the PC can access the temperature data (with a GET 101 instruction) and subtract it from a set-point value.

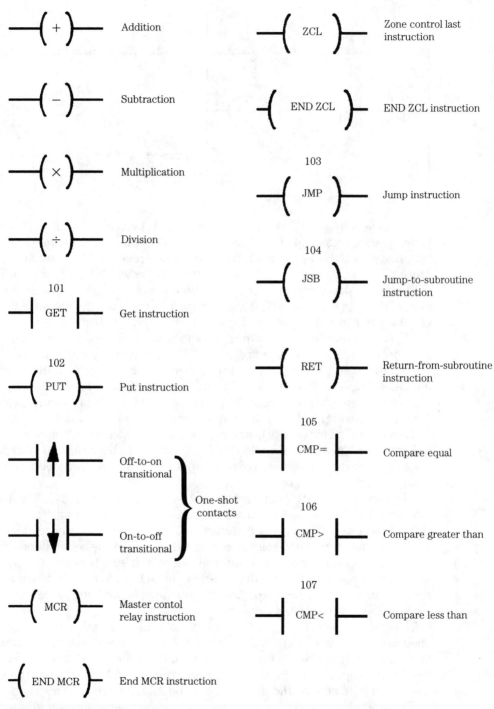

—(+)—   Addition

—(−)—   Subtraction

—(×)—   Multiplication

—(÷)—   Division

101
—| GET |—   Get instruction

102
—(PUT)—   Put instruction

—|↑|—   Off-to-on transitional

—|↓|—   On-to-off transitional

} One-shot contacts

—(MCR)—   Master contol relay instruction

—(END MCR)—   End MCR instruction

—(ZCL)—   Zone control last instruction

—(END ZCL)—   END ZCL instruction

103
—(JMP)—   Jump instruction

104
—(JSB)—   Jump-to-subroutine instruction

—(RET)—   Return-from-subroutine instruction

105
—| CMP= |—   Compare equal

106
—| CMP> |—   Compare greater than

107
—| CMP< |—   Compare less than

Figure 9.4 Extended programming symbols.

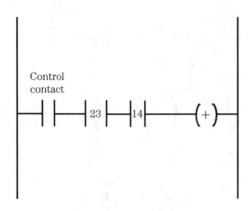

Figure 9.5 Addition of two constants.

These memory locations for the temporary storage of data, instructions, and information are often called *registers*. A PC register normally contains space for 16 bits; that is, a 16-bit number can be stored in one register. The manufacturer's literature should be consulted to determine the register width for a particular PC.

Figure 9.6 demonstrates the subtraction of data. When the control contact is closed, the data in registers 760 and 202 are accessed, and the datum in register 202 is subtracted from the datum in register 760. The number above the subtraction symbol is the register number in which the result is stored. In this case, the difference between register 760 and register 202 is stored in register 007.

Operations on both data and constants can also be performed. In Figure 9.7, the contents of register 702 are multiplied by the constant 10. The result is stored in register 521. When the control contact for rung 2 is closed, this product is divided by the contents of register 221, and the quotient is stored in register 336.

The companion to the GET instruction is the PUT instruction. This symbol is shown in Figure 9.4. The PUT instruction includes a number above the coil symbol, as does the GET instruction. It is used to store outputs in a particular register or to move data from one register to another.

Two very useful instructions are the *one-shot contacts* shown in Figure 9.4. The off-to-on transitional contact closes for one scan of the program when a triggering signal is received. The on-to-off transitional contact, on the other hand, opens for one program scan when the signal is received. Note that the contact does not remain closed (or opened, for the on-to-off transitional); it simply closes for one program scan. A typical application for the one-shot contact would be in unlatching a latched relay when a signal is received. Any operation performed only once will very likely employ the one-shot instruction.

The *master control relay* (MCR) and END MCR instructions shown in Figure 9.4 are used to control the operation of the program. The MCR and END MCR statements isolate a portion of the ladder program, much in the same way that a subroutine is isolated in a computer program. In Figure 9.8, rungs 4 and 5 constitute the portion of the program isolated by the MCR (rung 3) and END MCR (rung 6) instructions.

The program works as follows. The arithmetical operations of multiplication and division are performed by rungs 1 and 2. If control contact C3 is not closed, rungs 3 through 6 are automatically skipped, or ignored, and the program continues with rung 7. When the control contact for rung 7 is closed, timer 070 begins timing in 1-second intervals (BASE 01) until a preset value of 60 seconds (PRE 60) is obtained. At this point, counter 121 counts up by one unit (UP 01) from its preset value of zero (PRE 00). After the counting operation in rung 8 is performed, timer 070 is reset and once again times 1-second intervals up to a preset value of 60. If, however, control contact C3 is closed, rungs 4 and 5 are performed. Rung 6 then indicates the end of the isolated rungs, and the program proceeds to rung 7.

The group of ladder rungs controlled by the MCR and END MCR instructions is called a *zone* of ladder rungs. The MCR instruction is often used as an *override* instruction. Typically, an MCR/END MCR pair is included in a program to protect the PC in the event of an emergency. For example, in Figure 9.8, control contact C3 might close when current consumption by the power supply approaches a critical value. One-shot 4 would then unlatch relay 333, thereby protecting the system. The unlatching of relay 333 would trigger the normally closed contact in rung 5, and a critical piece of data would be obtained from register 336 and stored safely in register 402. The MCR instruction and the ZCL instruction (discussed next) are often called *skip* instructions.

The *zone control last state* (ZCL) and END ZCL instructions are almost identical to the MCR and END MCR instructions, respectively. The only significant difference is that if the MCR output is not activated (i.e., if the control contact for the MCR rung is not closed), the outputs within the MCR-controlled zone are de-energized. If the ZCL output is de-energized (i.e., if the control contact for the ZCL rung is not closed), the outputs within the ZCL-controlled zone are held in their last states.

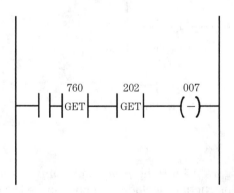

Figure 9.6 Subtraction of data.

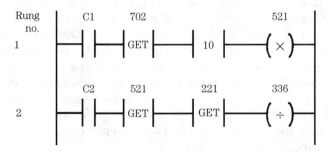

Figure 9.7 Arithmetic operations on data and constants.

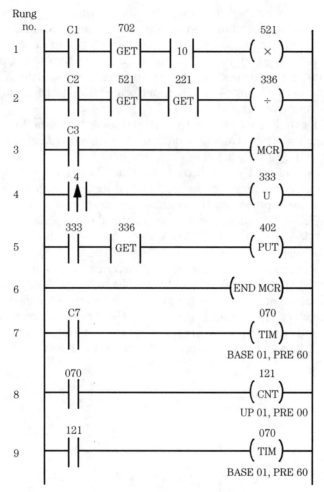

Figure 9.8 The MCR and END MCR instructions.

The next two instructions in Figure 9.4 are the *jump* and *jump-to-subroutine* (JSB) instructions. Both of these instructions include a number above the coil symbol that indicates the contact that is the object of the instruction. When the control contact for a rung containing a jump instruction is closed, the microprocessor is instructed to jump to the program rung containing the control contact referenced by the jump instruction. The jump instruction is thus a method of changing the order in which a program is executed. The jump-to-subroutine instruction is equivalent to the GOSUB instruction often found in computer programs. For example, when the control contact for the ladder rung containing the jump-to-subroutine instruction is closed, the microprocessor automatically jumps to the rung containing the control contact referenced in the JSB instruction. The microprocessor then proceeds to process the subroutine.

The end of the subroutine is marked with the RET instruction (also shown in Figure 9.4). RET stands for return-from-subroutine and directs the microprocessor to return to the program rung immediately following the JSB instruction. This procedure is demonstrated in Figure 9.9.

In Figure 9.9, the subroutine consists of ladder rungs 100 and 101. When the JSB instruction is encountered in rung 34, program execution immediately proceeds with rung 100. When the subroutine is executed and rung 102 is encountered, the microprocessor automatically returns to rung 35 to continue with the execution of the main program.

The three remaining symbols in Figure 9.4 are comparison operations. Note that all three symbols include numbers above them that refer to particular registers. These instructions are used in conjunction with GET instructions and are demonstrated in Figure 9.10. When the control contact in the first rung of the program is closed, the data in register 111 are compared with the data in register 105. If the two are equal, motor 1 is turned on. When the control contact in rung 2 is activated, data in register 112 are compared with data in register 106. If the value of the data in register 106 is greater than the value of the data in register 112, motor 2 is turned on. Similarly, when the control contact in rung 3 is closed, data in register 113 are compared with data in register 107. If the value of the data in register 107 is less than the value of the data in register 113, motor 3 is turned on.

Many more ladder symbols are used in the programming of PCs. The ones presented above are the symbols most often encountered in PC programming. Of course, not

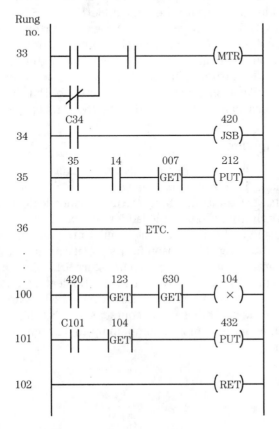

Figure 9.9 The JSB and RET instructions.

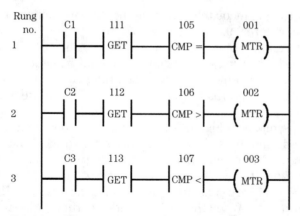

Figure 9.10 The CMP =, CMP>, and CMP< instructions.

every manufacturer uses the same symbols, as shown in Figure 9.11, which is the list of programming instructions for the Toshiba EX100 PC. Differences between the symbols presented in Figure 9.4 and the symbols presented in Figure 9.11 are apparent. Although the symbols are different, reading the supplier's user's manual is usually sufficient to highlight the meaning of each of the individual programming symbols.

Variations exist in the basic relay ladder programming language. One such variation is STEPS™, a relay ladder language for the StepLadder™ PLC. STEPS™ is discussed in detail in appendix C. Figure 9.12 shows a program written in STEPS™, which operates the domelight and door-ajar warning light in a two-door vehicle.

As noted in the Introduction, PC manufacturers have become better at producing readable, instructive manuals. Entertron Industries has produced an excellent monograph on relay ladder logic, available on floppy disk. The address for Entertron is given in appendix A. Ask for the user's manual for the SK1600-SK1800 on floppy disk.

Low-Level Languages: Boolean Language

Boolean language is based on the same principles as Boolean algebra, discussed in chapter 3. Boolean language is a *mnemonic* language, which means that the Boolean operators (AND, OR, NOT, NAND, and NOR) are used symbolically to provide a programming language entirely equivalent to the ladder language.

Boolean mnemonic language is the second most popular programming language, behind ladder-diagram programming. The reason for its popularity is that it is interchangeable with ladder programming, as can be seen in Figure 9.13. The familiar ladder diagram symbols in Figure 9.13 are equivalent to the Boolean symbols, which comprise two to six characters. In many cases, the Boolean characters are the same as the ladder symbols, minus the parentheses and contact symbols. This similarity is especially true of the timer, counter, compare, jump, jump-to-subroutine, and master control relay instructions.

Basic ladder functions

Instruction	Expression	Description	No. of steps
NO contact	─┤ ├─ ⒜	NO contact of device ⒜.	1
NC contact	─┤/├─ ⒜	NC contact of device ⒜.	1
Coil	─( )─ ⒜	Relay coil of device ⒜.	1
Forced coil	✕─( )─ ⒜	Forced coil of device ⒜. (The coil is specified as forced.)	1
Transitional contact (rising)	Input ─┤↑├─ Output ⒜	Turns ON output for 1 scan when input changes from OFF to ON.	1
Transitional contact (falling)	Input ─┤↓├─ Output ⒜	Turns ON output for 1 scan when input changes from ON to OFF.	1
Master control	Input ─[MCS]─ ⌇ ─[MCR]─	Turns OFF the power rail between MCS and MCR when MCS is OFF.	MCS, MCR 1 each
Jump control	Input ─[JCS]─ ⌇ ─[JCR]─	Jumps from JCS to JCR when JCS is ON.	JCS, JCR 1 each
ON delay timer	Input ─[⒜ TON ⒝]─ Output	Turns ON output when the time specified by ⒜ has elapsed after the input turns ON. (⒝ is a timer register.)	2/3
OFF delay timer	Input ─[⒜ TOF ⒝]─ Output	Turns OFF output when the time specified by ⒜ has elapsed after the input turns OFF. (⒝ is a timer register.)	2/3
Single-shot timer	Input ─[⒜ SS ⒝]─ Output	Turns ON output for the time specified by ⒜ when the input turns ON. (⒝ is a timer register.)	2/3
Counter	Count input ─[CNT ⒜ ⒝]─ Output Enable input	Counts the number of cycles the count input is ON while the enable input is ON and turns output ON when the number of cycles specified by ⒜ is reached. (⒝ is a counter register.)	2/3
End	─[END]─	Specifies the end of a program.	1

Figure 9.11 Programming instruction for the Toshiba EX100 PC (reprinted with permission of Toshiba International Corporation).

Data transfer instructions

Instruction (FUN. No.)	Expression	Description	No. of steps
Register transfer (FUN. 000)	Ⓐ W → W Ⓑ	Transfers data from register Ⓐ to register Ⓑ.	3
Constant transfer (FUN. 001)	Ⓐ K → W Ⓑ	Transfers 16-bit constant Ⓐ to register Ⓑ.	3 / 4
Table initialization (FUN. 002)	Ⓐ TINZ [nn] Ⓑ	Transfers data from register Ⓐ to every register in the table, size [nn] starting with register Ⓑ.	4
Multiplexer (FUN. 003)	Ⓐ T → W [nn] Ⓑ → Ⓒ	Transfers data from the register specified by Ⓑ in the table, size [nn] starting with register Ⓐ, to register Ⓒ.	5
Demultiplexer (FUN. 004)	Ⓐ W → T [nn] Ⓑ → Ⓒ	Transfers data from register Ⓐ to the register specified by Ⓑ in the table, size [nn] starting with register Ⓒ.	5
Table block transfer (FUN. 005)	Ⓐ T → T [nn] Ⓑ	Transfers the data of every register in the table, size [nn] starting with Ⓐ, to the registers after Ⓑ.	4

Arithmetic operations

Instruction (FUN. No.)	Expression	Description	No. of steps
Register addition (FUN. 010)	Ⓐ + Ⓑ → Ⓒ	Adds the data in registers Ⓐ and Ⓑ and stores the result in register Ⓒ.	4
Register subtraction (FUN. 011)	Ⓐ - Ⓑ → Ⓒ	Subtracts register Ⓑ from register Ⓐ and stores the result in register Ⓒ.	4
Register multiplication (FUN. 012)	Ⓐ × Ⓑ → Ⓒ	Multiplies the data in registers Ⓐ and Ⓑ and stores the result in double-length register Ⓒ · Ⓒ + 1.	4
Register division (FUN. 013)	Ⓐ / Ⓑ → Ⓒ	Divides the data in double-length register Ⓐ · Ⓐ + 1 by register Ⓑ and stores the quotient in Ⓒ and the remainder in Ⓒ + 1.	4
Register comparison (FUN. 014)	Ⓐ > Ⓑ ├ Output	Compares registers Ⓐ and Ⓑ and turns on the output if the result is true.	3
Register comparison (FUN. 015)	Ⓐ = Ⓑ ├ Output	Compares registers Ⓐ and Ⓑ and turns on the output if the result is true.	3
Register comparison (FUN. 016)	Ⓐ < Ⓑ ├ Output	Compares registers Ⓐ and Ⓑ and turns on the output if the result is true.	3

Figure 9.11 Continued.

Arithmetic operations (cont'd)

Instruction (FUN. No.)	Expression	Description	No. of steps
Double-length addition (FUN. 017)	Ⓐ + + Ⓑ → → Ⓒ	Adds the data of double-length (32-bit) registers Ⓐ · Ⓐ + 1 and Ⓑ · Ⓑ + 1 and stores the result in Ⓒ · Ⓒ + 1.	4
Double-length subtraction (FUN. 018)	Ⓐ − − Ⓑ → → Ⓒ	Subtracts the data of double-length registers Ⓑ · Ⓑ + 1 from Ⓐ · Ⓐ + 1 and stores the result in Ⓒ · Ⓒ + 1.	4
Constant addition (FUN. 020)	Ⓐ + . Ⓑ → Ⓒ	Adds the data in register Ⓐ and constant Ⓑ and stores the result in Ⓒ.	4 / 5
Constant subtraction (FUN. 021)	Ⓐ − . Ⓑ → Ⓒ	Subtracts constant Ⓑ from the data in register Ⓐ and stores the result in Ⓒ.	4 / 5
Constant multiplication (FUN. 022)	Ⓐ × . Ⓑ → Ⓒ	Multiplies constant Ⓑ with the data in register Ⓐ and stores the result in double-length register Ⓒ · Ⓒ + 1.	4 / 5
Constant division (FUN. 023)	Ⓐ / . Ⓑ → Ⓒ	Divides the data in double-length register Ⓐ · Ⓐ + 1 by constant Ⓑ and stores the quotient in Ⓒ, and the remainder in Ⓒ + 1.	4 / 5
Constant comparison (FUN. 024)	Ⓐ > . Ⓑ ├ Output	Compares the data in register Ⓐ with constant Ⓑ and turns on the output if the result is true.	3 / 4
Constant comparison (FUN. 025)	Ⓐ = . Ⓑ ├ Output	Compares the data in register Ⓐ with constant Ⓑ and turns on the output if the result is true.	3 / 4
Constant comparison (FUN. 026)	Ⓐ < . Ⓑ ├ Output	Compares the data in register Ⓐ with constant Ⓑ and turns on the output if the result is true.	3 / 4

Logical operations

Instruction (FUN.No.)	Expression	Description	No. of steps
Register AND (FUN. 030)	Ⓐ AND Ⓑ → Ⓒ	ANDs the data of registers Ⓐ and Ⓑ and stores the result in register Ⓒ.	4
Register OR (FUN. 031)	Ⓐ OR Ⓑ → Ⓒ	ORs the data of registers Ⓐ and Ⓑ and stores the result in register Ⓒ.	4
Register exclusive OR (FUN. 032)	Ⓐ EOR Ⓑ → Ⓒ	Exclusive ORs the data in registers Ⓐ and Ⓑ and stores the result in register Ⓒ.	4
Register inversion (FUN. 033)	Ⓐ NOT Ⓑ	Inverts each bit of register Ⓐ and stores the result in register Ⓑ.	3
Right rotation (FUN. 035)	Ⓐ RTR Ⓑ → Ⓒ	Rotates the bits in register Ⓐ to the right by the number specified in register Ⓑ and stores the result in register Ⓒ.	4

Figure 9.11 Continued.

Logical operations

Instruction (FUN.No.)	Expression	Description	No. of steps
Left rotation (FUN. 036)	Ⓐ RTL Ⓑ → ©	Rotates the bits in register Ⓐ to the left by the number specified in register Ⓑ and stores the result in register ©.	4
Constant AND (FUN. 040)	ⒶAND.Ⓑ → ©	ANDs the data in register Ⓐ and constant Ⓑ and stores the result in register ©.	4 / 5
Constant OR (FUN. 041)	Ⓐ OR.Ⓑ → ©	ORs the data in register Ⓐ and constant Ⓑ and stores the result in register ©.	4 / 5
Constant exclusive OR (FUN. 042)	Ⓐ EOR.Ⓑ → ©	Exclusive ORs the data in register Ⓐ and constant Ⓑ and stores the result in register ©.	4 / 5
Bit test (FUN. 043)	Ⓐ TEST Ⓑ — Output	ANDs the data in register Ⓐ and constant Ⓑ and turns on the output if the result is other than 0.	3 / 4
Complemen (FUN. 046)	Ⓐ NEG Ⓑ	Calculates the 2's complement of data in register Ⓐ and stores the result in Ⓑ.	3

Data conversion instructions

Instruction (FUN. No.)	Expression	Description	No. of steps
Binary conversion (FUN. 050)	Ⓐ BIN Ⓑ	Converts the BCD data in register Ⓐ into binary, and stores the result in register Ⓑ.	3
Single-length BCD conversion (FUN. 051)	Ⓐ BCD1 Ⓑ	Converts the binary data in register Ⓐ into 4-digit BCD code and stores the result in register Ⓑ.	3
Double-length BCD conversion (FUN. 052)	Ⓐ BCD2 Ⓑ	Converts the binary data in double-length registers Ⓐ · Ⓐ + 1 into BCD code and stores the result Ⓑ · Ⓑ + 1 · Ⓑ + 2.	3
Encode (FUN. 053)	Ⓐ ENC Ⓑ	Converts the most significant ON bit position of Ⓐ into 4-bit data and stores the result in Ⓑ.	3
Decode (FUN. 054)	Ⓐ DEC Ⓑ	Converts the least significant 4-bit data of Ⓐ into bit position and stores the result in Ⓑ.	3
Bit count (FUN. 055)	Ⓐ BITC Ⓑ	Counts the number of bits that are ON in register Ⓐ and stores the result in register Ⓑ.	3

Figure 9.11 Continued.

Special
functions

Instruction (FUN. No.)	Expression	Description	No. of steps
Upper limit (FUN. 060)	Ⓐ UL Ⓑ → Ⓒ	Sets the data in register Ⓐ as the upper limit according to register Ⓑ and stores the result in register Ⓒ.	4
Lower limit (FUN. 061)	Ⓐ LL Ⓑ → Ⓒ	Sets the data in register Ⓐ as the lower limit according to register Ⓑ and stores the result in register Ⓒ.	4
Maximum value (FUN. 062)	Ⓐ MAX [nn] Ⓑ	Locates the maximum value in table size [nn] starting with register Ⓐ and stores it in register Ⓑ. Stores the pointer indicating the value in register Ⓑ + 1.	4
Minimum value (FUN. 063)	Ⓐ MIN [nn] Ⓑ	Locates the minimum value in table size [nn] starting with register Ⓐ and stores it in register Ⓑ. Stores the pointer indicating the value in register Ⓑ + 1.	4
Average value (FUN. 064)	Ⓐ AVE [nn] Ⓑ	Calculates the average value of table size [nn] starting with register Ⓐ and stores it in register Ⓑ.	4
Function generator (FUN. 065)	Ⓐ FG [nn] Ⓑ → Ⓒ	Generates optional functions by registering function parameters.	5
Square root (FUN. 070)	Ⓐ RT Ⓑ	Calculates the square root of the data in double-length register Ⓐ · Ⓐ + 1 and stores the result in register Ⓑ.	3
Sine function (FUN. 071)	Ⓐ SIN Ⓑ	Divides the data in register Ⓐ by 100 and obtains its sine. Mutiplies the answer by 10000 and stores the result in register Ⓑ.	3
Arcsine function (FUN. 072)	Ⓐ ASIN Ⓑ	Divides the data in register Ⓐ by 10000 and obtains its arcsine. Mutiplies the answer by 100 and stores the result in register Ⓑ.	3
Cosine function (FUN. 073)	Ⓐ COS Ⓑ	Divides the data in register Ⓐ by 100 and obtains its cosine. Mutiplies the answer by 10000 and stores the result in register Ⓑ.	3
Arccosine function (FUN. 074)	Ⓐ ACOS Ⓑ	Divides the data in register Ⓐ by 10000 and obtains its arccosine. Mutiplies the answer by 100 and stores the result in register Ⓑ.	3

Figure 9.11 Continued.

Other functions

Instruction (FUN. No.)	Expression	Description	No. of steps
Device set (FUN. 080)	SET Ⓐ	Turns on device Ⓐ.	2
Device reset (FUN. 081)	RST Ⓐ	Turns off device Ⓐ.	2
Diagnostic display (FUN. 080)	DDSP Ⓐ	When the input turns ON, the diagnostic code set by Ⓐ is displayed on a peripheral device.	2 / 3
Diagnostic display with message (FUN. 091)	DDSM Ⓐ Ⓑ	When the input turns ON, the diagnostic code set by Ⓐ and the message registered in the registers after Ⓑ are displayed on a peripheral device.	3 / 4
Immediate input (FUN. 096)	IN [nn] Ⓐ	Immediately updates the input data of [nn] registers starting with Ⓐ.	3
Immediate output (FUN. 097)	OUT [nn] Ⓐ	Immediately updates the output data of [nn] registers starting with Ⓐ.	3
Step sequence initialization (FUN. 100)	STIZ [nn] Ⓐ	Initializes the step sequencer beginning with device Ⓐ.	3
Step sequence input (FUN. 101)		Sets the step sequencer suitable for sequential control.	2
Step sequence output (FUN. 102)	Ⓐ		2
Flip-flop (FUN. 110)	Set input — F / F Reset input — Ⓐ	Turns on device Ⓐ when the set input turns on and turns off device Ⓐ when the reset input turns on.	2
Up/down counter (FUN. 111)	Up / down Select input — U / D Count input Enable input — Ⓐ	While the enable input is on, counts up or down the number of cycles the count input turns on, according to the up / down select input. Select input: ON = up, OFF = down	2
Shift register (FUN. 112)	Data input — SR Shift input — [nn] Enable input — Ⓐ	If the shift input is on while the enable input is on, shifts the data in the shift register by one bit. Shift register: [nn] bits of data starting with device Ⓐ.	3

Figure 9.11 Continued.

STEPS Program - domelite   date - 03-29-1991   Time - 16:35:38

```
**** Program DOMELITE  ***** For The StepLadder PLC ***

Written with STEPS programming, Domelite operates the dome light and the
door ajar warning light in a two door vehicle.  This program uses 36 bytes
of the 512 available in The StepLadder PLC.
     Lines starting with '@', assign names to input or output points.
Lines that start with '<', define logical seqences and end with an output.
The period '.' means AND, and the coma ',' means OR.

@DoorA=2
@DoorB=3
@SwDoor=1
@SwOn=4
@DomeLite=0
@DoorAjar=5
@Buzz=6

<SwDoor.(\DoorA,\DoorB),SwOn>DomeLite
     This is a logic line to turn the dome light on or off.
     STEPS can draw the logic circuit above the logic line.
<\DoorA,\DoorB>DoorAjar
     This logic line turns the door ajar warning light on or off.
<DoorAjar>Buzz
     Finally a line to turn on the buzzer
```

Figure 9.12 DOMELITE program written in STEPS ™ (reprinted with permission of Active Systems Group,

Five of the symbols in Figure 9.13 deserve special mention. These are the OR, NOR, OUT NOT, LOAD, and LOAD NOT instructions. The OR instruction is simply an N.O. contact in parallel with a rung. Continuity through the rung can be achieved by closing the contacts in the rung, or by closing the parallel contact. The NOR (or NOT OR) instruction is similar, except that it employs an N.C. contact. The OUT NOT is similar to the OUT symbol, with one exception. An OUT symbol is energized when continuity through the rung is achieved. An OUT NOT symbol is de-energized when continuity through the rung is achieved. The LOAD and LOAD NOT instructions are used to symbolize the initiation of the rung, with N.O. or N.C. control contacts, respectively.

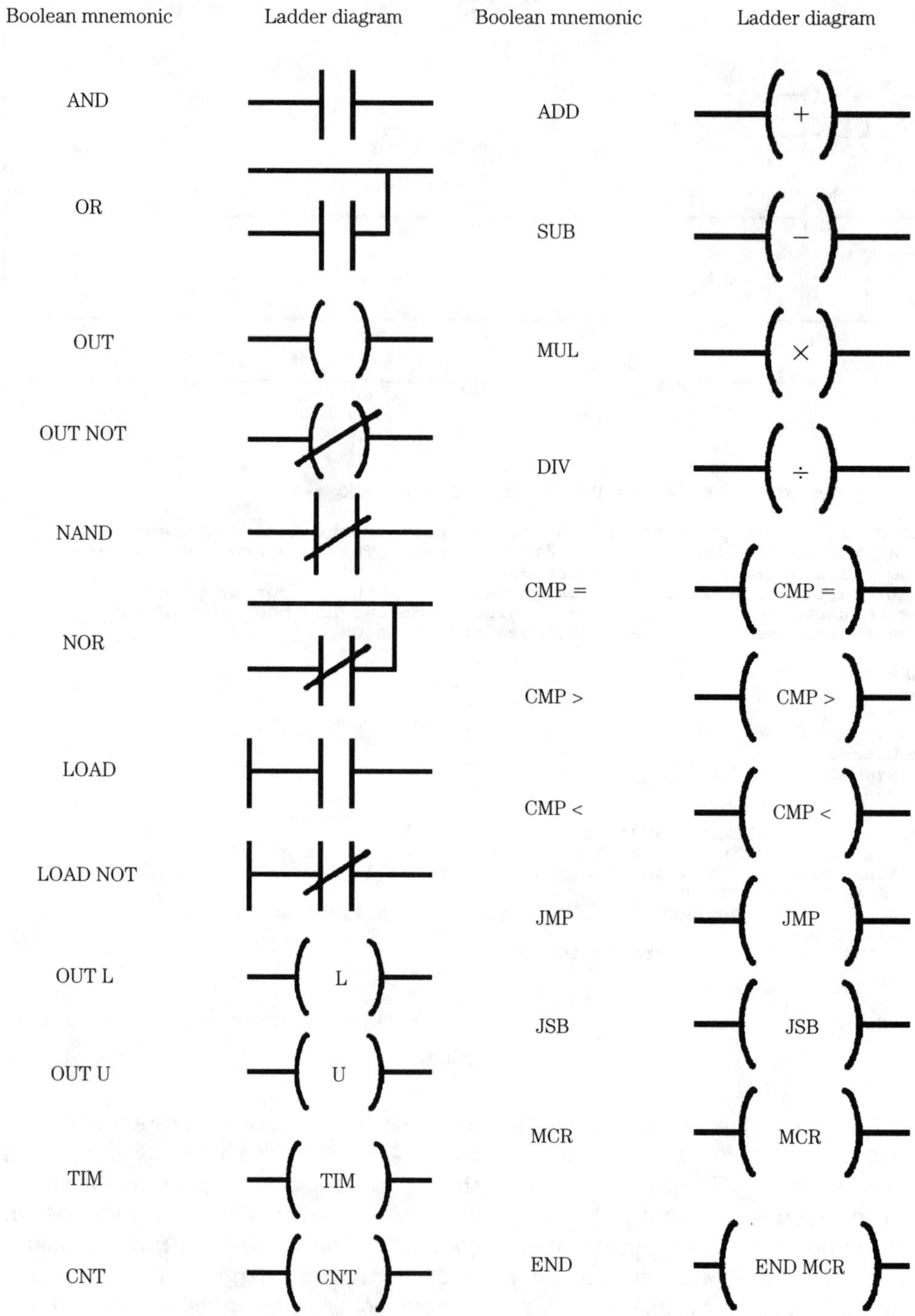

Boolean mnemonic	Ladder diagram	Boolean mnemonic	Ladder diagram
AND		ADD	
OR		SUB	
OUT		MUL	
OUT NOT		DIV	
NAND		CMP =	
NOR		CMP >	
LOAD		CMP <	
LOAD NOT		JMP	
OUT L		JSB	
OUT U		MCR	
TIM		MCR	
CNT		END	END MCR

Figure 9.13 Comparison of Boolean mnemonic and ladder diagram languages.

High-Level Languages: Block Diagrams

A low-level language, such as the ladder diagram or Boolean mnemonic symbol languages discussed in the previous sections, has limited applicability when programming complex routines. To implement fairly complex programs (programs that, for example, include proportional-integral-derivative control), a high-level language is needed. Block diagrams can be used to form a high-level language that allows the programmer to implement complex routines using the ladder-diagram format.

The block-diagram symbols most often encountered are shown in Figure 9.14. These symbols are used in conjunction with the ladder symbols for the N.O. and N.C. contacts and the OUT and OUT NOT output conditions. Block diagram language uses the ladder format; that is, the block symbols are placed in rungs and require energizing signals.

The first two symbols in Figure 9.14, the timer and counter block symbols, illustrate the general form that block diagrams assume. Two inputs are shown. The control input is in series with a control contact. When the control contact is closed, the timer begins timing and the counter begins counting. Within the block, the preset and base values are shown. The counter includes a direction; in this case, the counter counts down from a preset value of 30 in increments of 1 each time the control contact is closed. The base and preset values have the same meaning as in Figure 9.8. For example, when the control contact in the timer rung is closed, the timer begins timing in 1-second intervals up to the preset value of 10 seconds. At 10 seconds, the output goes ON or TRUE (i.e., the output is energized), and continuity through the rung is assured.

The reset inputs are not always shown in block symbols. For example, some timer symbols do not include a reset input. When the reset input is not present in the timer symbol, it is assumed that the timer automatically resets to begin a subsequent timing cycle when its control contact is closed.

It should be emphasized that the counter in Figure 9.14 counts down from the preset value in increments of 1. In other words, when the control contact is closed, the count value drops to 29. The count value remains at 29 *until the control contact is again closed*. To count down from 30 to 0, the control contact must be closed 30 times. Some counter symbols include a second output line that indicates the running count value.

The third symbol in Figure 9.14 is the block language symbol for arithmetical operations (addition, subtraction, multiplication, and division). When the control contact is closed (and the control input is energized), the number contained in register 107 (which can be either data or a constant) and the number contained in register 302 are used to perform an arithmetical operation. If the block symbolizes addition, the two numbers are added. If the block symbolizes subtraction, the number in register 302 is subtracted from the number in register 107. The same holds true for multiplication and division. (In the case of division, the number in register 107 is divided by the number in register 302.) The result of the operation (the sum, difference, product, or quotient) is then stored in register 416.

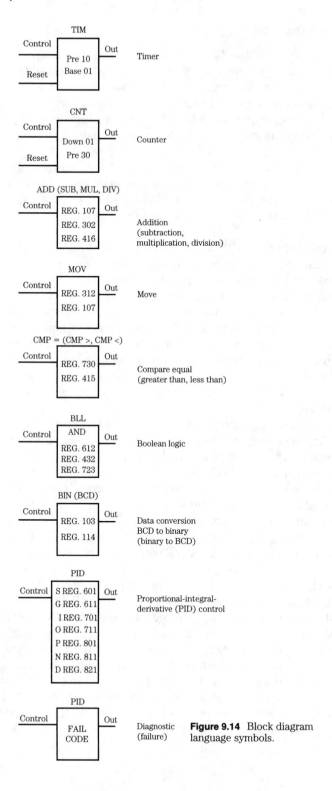

Figure 9.14 Block diagram language symbols.

The output line to the right of the arithmetical operation block is not really needed because the result of the operation is stored in a register location. The output line is generally used as an *overflow indicator*. Whenever the result of the operation is greater than the register can hold, the output line is energized. If a warning light is connected to the output, the light signals the programmer that the capacity of the register has been exceeded. A typical overflow condition occurs when two eight-bit numbers are added to produce a nine-bit sum in a register that is only eight bits wide. This type of overflow can be corrected by a process known as *scaling*. The specific PC's user's manual should be consulted for details on the scaling process.

The *move* instruction in Figure 9.14 is the equivalent of the ladder instructions GET and PUT. When the control contact is closed, the number in register 312 is acquired (GET) and moved (PUT) to register 107. When this operation is completed, the output is energized.

Note that the Toshiba instruction set contains another data transfer instruction, the *shift* register. The shift register instruction simply shifts words within a stack of registers each time an operation is performed. (Register stacks are found in the data section of application memory; refer to chapter 6.)

The three compare instructions (CMP =, CMP >, CMP <) can be represented in block form (Figure 9.14). Numbers in the two registers are compared, and if the equality or inequality condition is satisfied, the output is energized.

The Boolean logic (BLL) block symbol performs the Boolean operations discussed in chapters 2 and 3: AND, OR, NOT, NAND, NOR, and XOR. The operation to be performed is specified within the block. Thus, when the control input is ON, the symbol shown in Figure 9.14 performs the AND operation on registers 612 and 432, and stores the result in register 723. The output is energized when the operation is performed.

Not all PCs have sufficient memory capacity to store the results of block operations in separate registers. If separate registers do not exist, the result of a block operation is stored in one of the input registers, thereby erasing the input value. For example, if the BLL block in Figure 9.14 contains only registers 612 and 432, the result of the AND operation would be stored in register 432. The data previously stored there would be erased.

The three remaining blocks shown in Figure 9.14 have no analogs in either the ladder or Boolean mnemonic languages and demonstrate the advantage of a high-level language. These symbols are data conversion, proportional-integral-derivative (PID) control, and diagnostic.

Although the data conversion block in Figure 9.14 shows the conversion of binary-coded decimal (BCD) into binary (or binary into BCD), other conversions can occur. For example, it can include the conversion of 16-bit data into 8-bit data for seven-segment displays and the complement, which inverts data, in addition to the BCD-to-binary and binary-to-BCD instructions, must be stored. The operation of the block in Figure 9.14 is simple. When the control input is ON, BCD data in register 103 are converted to binary and stored in register 114. When this operation is completed, the output turns ON.

Proportional-integral-derivative (PID) control was discussed in chapter 7 under "Special I/O interfaces." That chapter emphasized the use of intelligent I/O interfaces to perform PID control. PID control can also be programmed into the PC's memory using the symbol shown in Figure 9.14.

The set point for the process (S) and desired-gain (G) values are entered into registers 601 and 611, respectively. Input from the process (I) and the output control signal (O) use registers 701 and 711, respectively. The three remaining registers (801, 811, 821) hold the proportional (P), integral (N), and derivative (D) terms, respectively. These terms comprise the PID control signal. When the control line is ON, the output line carries the PID control signal.

The absence of a PID programming symbol from an instruction set does not necessarily mean that the PC cannot perform PID control. PID is a popular control scheme, and many PC manufacturers include the PID scheme in a module connected to the PC. In such a case, PID control is initiated with a simple ladder output instruction in the main program.

The diagnostic block shown in Figure 9.14 is quite useful. When energized, this particular diagnostic symbol indicates, via a failure code, the type or mode of failure that has occurred. If, for example, an interruption occurs in the power supply to a portion of the system, the diagnostic program is energized. The program then determines that the type of failure was a power interrupt and sends an output signal (failure code) to an output display (such as two seven-segment displays) in the diagnostic block's output line. The two-digit number displayed by the seven-segment displays alerts the operator that a failure has occurred. The operator compares the two-digit number to a list of failure codes and determines that a power interrupt was the type of failure suffered. A wide variety of diagnostics are available with various PCs. The PC's user's manual should be consulted for further details.

High-Level Languages: Computer-Type Languages

Users of personal computers will be familiar with the types of high-level languages discussed in this section. *Computer-type languages* (CTL) are languages that employ English statements and instructions. They are usually similar to BASIC, the widely used personal computer programming language. In fact, several PC manufacturers use BASIC as their programming language.

Although similar to BASIC, most CTLs are easier to use than BASIC; i.e., they are more *user-friendly* or *operator-oriented*. Because CTLs essentially offer the complete programming flexibility of a computer language, their use enhances the computational and control power of the PC. And, because CTLs are based on the English language, they are easier for the operators to understand.

The specifics of a particular CTL can be obtained from the PC's user's manual. Because most CTLs are quite similar to BASIC, we examine the BASIC instruction set, which is shown in Table 9.1. Although these instructions do not correspond exactly to ladder symbols, they do display similarities. The first four instructions, LET, INPUT, READ, and DATA, are similar in nature to the ladder contact symbols. The LET instruction is used to assign a number value to a variable. Thus, if a contact is represented by the variable X, the statement LET X = 1 indicates the contact is closed. The instruction INPUT accepts data from the PC's keyboard. A program containing the statement INPUT X causes a prompt (such as a question mark) to be displayed on the terminal screen (or CRT). The program will not continue until the value of X is entered. The READ instruction is similar to the INPUT instruction, except that it directs the microprocessor to accept data from inputs to the PC or from DATA statements. For example, READ T directs the microprocessor to accept a temperature input from a thermocouple. A DATA statement is a line in a program that contains data to be used in the execution of the program (e.g., DATA 73.46, 892.10, 107.62). Data in the DATA statement are accessed by the READ instruction.

TABLE 9.1

LET	IF	TIMER ON	WAIT
INPUT	THEN	TIMER OFF	FOR
READ		TIMER STOP	STEP
DATA			NEXT
+	PEEK	GOTO	=
-	POKE	GOSUB	>
*		RETURN	<
/			
SQR			
LOG			
SIN			
COS			
etc.			
ERL	PRINT	REM	END
ERR			RUN

The IF and THEN instructions are used together and simulate the output ladder symbol. Suppose that we wish to simulate a ladder rung that contains one N.O. contact (X) and one output (Y). When the N.O. contact is closed, the output is energized, which can be simulated using the following program:

```
10 READ X
20 IF X=0 THEN Y=0
30 IF X=1 THEN Y=1
```

In other words, the IF/THEN pair establishes a condition that, when satisfied, results in a specific output. Note that each line in the program given above is numbered for programming convenience.

Timer operation is controlled by the TIMER ON, TIMER OFF, and TIMER STOP instructions. Timer instructions vary widely among different versions of BASIC. In our version, these instructions turn on, turn off, and temporarily halt the timer.

The WAIT, FOR, STEP, and NEXT instructions can be used to simulate the operation of a counter. The WAIT instruction halts execution of a program until specified inputs are provided. For example, suppose that we have written a program that controls the sealing of a carton. We do not want the program to begin functioning until 10 objects are placed in the carton. By using a light beam and photoelectric cell, we can obtain an electrical signal every time an object reaches the carton. If this input signal is connected to I/O terminal 16 on the PC, we can use the following line in the control program:

```
80 WAIT 16, 10
```

This line (line number 80 in the program) simply delays the execution of the remainder of the program until 10 signals have been received at I/O terminal 16.

The FOR, STEP, and NEXT instructions are less efficient in simulating counters, but they are very useful in other applications. FOR and STEP are used as a pair in the same program line. These three instructions allow an operation to be performed a preset number of times, and allow initial and final conditions to be set. A typical usage is as follows:

```
10 FOR X=0 TO 10 STEP 1
20 Y=2*X
30 PRINT "X=",X,"Y=",Y
40 NEXT X
```

In this program X is varied from zero to 10 in increments ("steps") of one (0, 1, 2, 3, . . ., 10). At each value of X, the value of Y (which is just two times X) is calculated, and the values of both X and Y are printed. The NEXT X instruction returns the program to line number 10 and increments the value of X by one unit.

The program given above uses the PRINT instruction, which is a very convenient method of obtaining output data from PCs. The output from the above program would appear as follows:

```
X = 0    Y = 0
X = 1    Y = 2
X = 2    Y = 4
X = 3    Y = 6
X = 4    Y = 8
X = 5    Y = 10
X = 6    Y = 12
X = 7    Y = 14
X = 8    Y = 16
X = 9    Y = 18
X = 10   Y = 20
```

In line 30 of the program, both X = and Y = are enclosed by quotation marks. Any data enclosed by quotation marks will be printed exactly as shown. Thus, in each cycle of program execution, X = and Y = are printed prior to the printing of the values of X and Y, respectively. The commas in line 30 direct the printer to allow some spaces between X = and the value of X, and Y = and the value of Y.

The next set of instructions (+, −, *, /, SQR, LOG, SIN, and COS) are mathematical instructions. They represent addition, subtraction, multiplication, division, square root extraction, logarithm computation, sine, and cosine computation, respectively. The instructions are used in equation form. Consider, for example, the following program:

```
10 INPUT U,V,W,X
20 LET G=[(U+V) * (W-X)]/(W*X)
```

```
30 LET Y = G + LOG(G) + SIN(G)/COS(G)
40 PRINT "G =", G, "Y =", Y
```

When this program is run, the values of U, V, W, and X are provided via the keyboard. The program then calculates G, which is given by the equation shown in line 20. Line 30 then adds G, its logarithm, and its tangent (tan(G) = sin(G)/cos(G)) together. Both G and Y are then printed.

The ease with which complex mathematical operations can be performed represents one of the great advantages of CTL over other types of programming languages. It should be noted that Boolean algebra can be performed with these mathematical instructions by replacing AND with *, OR with +, and so on. Chapter 3 should be consulted for further details.

The PEEK command followed by a register number or memory location functions the same as the GET ladder instruction followed by a register number or memory location. The POKE command, followed by a register number or memory location and a byte of data, functions the same as the PUT ladder instruction, which uses a register number or memory location. GOTO is similar to the JUMP ladder instruction. It directs the program to jump to a specified program line number. It is often used with the IF/THEN pair:

```
80 IF X = 72 THEN GOTO 150
```

The GOSUB instruction followed by a program line number and the RETURN instruction are identical to the JUMP-TO-SUBROUTINE and RETURN ladder instructions, respectively.

BASIC, like the ladder diagram, Boolean, and block languages, provides comparison instructions. These instructions are often used with the IF/THEN and GOTO instructions:

```
110 IF X = 0 THEN GOTO 200
120 IF X > 0 THEN GOTO 300
130 IF X < 0 THEN GOTO 140
```

In the program above, if X equals zero, the program skips ahead to line 200. If X is greater than zero, the program skips ahead to line 300. Finally, if X is less than zero, the program proceeds with line 140. (Note that line 130 in the program above is not necessary and is included only to demonstrate the less than comparison.)

The BASIC programming language also contains diagnostic instructions. The two shown in Table 9.1 (ERL and ERR) are extremely useful. ERL is used in conjunction with a print statement to indicate the number of the program line that contains an error:

```
110 PRINT ERL
```

ERR is used to identify the type of error that has occurred:

```
290 IF ERR = 26 THEN GOTO 500
```

In this example, error 26 indicates that a FOR instruction has been used without a subsequent NEXT instruction.

REM is a remark line instruction. If a program line is prefaced by REM, the contents of the line are not included as part of the program. REM instructions can be very helpful to operators:

```
10 REM THIS PROGRAM IS FOR CONTROL OF
20 REM TEMP IN STEAMLINES
30 INPUT TS
40 READ T
etc ...
```

The END and RUN instructions are necessary for the proper execution of a program. END must be present in all programs. It is the last (i.e., highest-numbered) line in the program. The program will not run without the END instruction:

```
990 PRINT "TS =", TS, "T =", T
1000 END
```

When RUN is typed into the keyboard and entered, program execution begins.

Although this section has concentrated on BASIC, there are a wide variety of CTL programming languages. One very interesting one, State Logic Control, is touted by Adatek, Inc. as being clearly superior to all other CTLs. Information about Adatek's State Logic Control can be found in appendix B.

Summary

Two principal types of languages are used in PC software: low-level and high-level. The two low-level languages discussed in this chapter were the ladder diagram and Boolean languages. The ladder diagram language is the most popular of the programming languages. It is the oldest PC language, and it is very easy for those who worked with the old relay-logic control systems to understand. Boolean, or mnemonic, language is the second most popular language. Both of these languages, however, are limited in flexibility.

Block diagrams and computer-type languages (CTLs) were two high-level programming languages discussed in this chapter. Both languages are more powerful and flexible than low-level languages. Block-diagram language is designed to fit the ladder diagram programming format and is thus popular among those familiar with the old relay-logic control systems. CTLs, such as BASIC, are quite popular among computer-literate operators because they are similar to the languages used to program personal computers. CTLs are the most powerful and flexible of all the PC programming languages.

PC Applications

Having examined the basics of programmable controller (PC) operations, the hardware, and the software associated with PCs, we turn to PC applications. Using PCs in industrial situations requires the operator to be familiar with PC programming. The first section of this chapter discusses the fundamentals of PC programming. The principles set forth in previous chapters are useful here. Subsequent sections of this chapter give examples of the utility of PCs in control applications.

Elements of Programming

GIGO is an acronym for Garbage In, Garbage Out. It is an acronym that should be kept in mind when programming the PC. After all, the PC can do only what it is instructed to do. If the instructions are inadequate or incorrect, the results will be unacceptable.

Before the control program, or *algorithm*, can be written, the programmer must understand clearly the nature of the task to be performed by the PC. It is often useful for the programmer to write out in clear English statements the control task to be performed and the method (e.g., PID) by which the task will be accomplished. For example, in chapter 5, a control scheme for controlling steam temperature was discussed. The English-language description of this control problem is as follows:

> The task is to control steam temperature in the line. The method for controlling it is the following: steam temperature will be measured downstream from the water injection site. When the temperature exceeds the set-point value, a pump will be activated. This pump will spray feedwater into the steam line at the water injection site, thereby cooling the steam. The control strategy will therefore be proportional control; i.e., the error signal will be proportional to the difference between actual steam temperature and the set point value.

Once the English-language description of the control problem has been written, the next step is to construct a *flow chart* of the control program. A flow chart is simply a schematic diagram of a program. It employs special symbols to denote the start, end, decision points, and so on of a program. Figure 10.1 shows the most commonly used flow chart symbols. The terminal point symbol indicates the beginning or end of a program. The process symbol is a schematic representation of program instructions that control program processing. The I/O point symbol indicates the presence of instructions that read input data or provide output signals. The decision point symbol indicates instructions in the program that require a decision to be made or that discriminate between logical states (equal to, greater than, or less than).

The flow chart for the steam temperature control program is shown in Figure 10.2. At the start of the program, the set-point temperature for the steam is entered, indicated by the process symbol. The program flows to the I/O point, which reads the actual temperature of the steam in the line. This temperature is then compared with the set-point temperature at the decision point. If the steam temperature in the line is not greater than the set-point temperature, the flow of the program reverts to the I/O point; in other words, the steam temperature is once again read. This series of program steps is repeated until the measured steam temperature is greater than the set-point temperature.

When the steam temperature exceeds the set-point temperature, the program flows from the decision point to another I/O point, which turns the feedwater pump on. This action lowers the steam temperature in the line. Note that the control task and the method of achieving the task are both clearly evident from an examination of the program's flow chart.

Once the flow chart has been drawn, the next step is to write the program. The relay ladder and contact diagrams for steam temperature control given in Figures 5.8 and 5.10 will not suffice, however, because those figures show the ladder program for the relay control systems of Figure 5.6. The PC setup for the control of steam

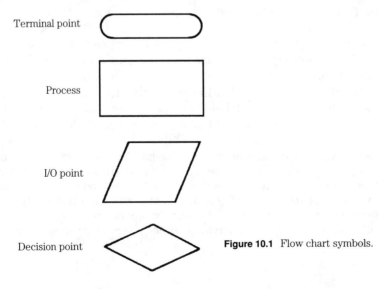

Terminal point

Process

I/O point

Decision point

Figure 10.1 Flow chart symbols.

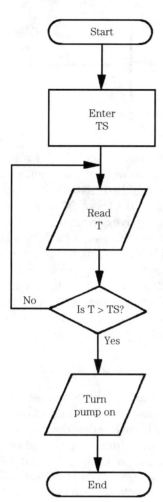

Figure 10.2 Flow chart for the steam temperature control program.

temperature is shown in Figure 7.2; it employs the ladder program shown in Figure 10.3. It can be seen that the program is quite simple. The set-point temperature is stored in register 112, and the output from the thermocouple in register 106. When the temperature exceeds the set-point value, the CMP > instruction goes ON, and the motor for the pump is activated.

Once the program is written, it must be entered into the PC. This can be done with the programming devices discussed in chapter 8. The choice of a programming device is governed by the make and model of the PC being used. (The PC's user's manual should be consulted for details.)

Two steps remain in programming a PC. The program should be *debugged*, which is a sometimes tedious process of locating and correcting mistakes so that the proper control function is performed. This step in the process is made easier if sufficient care is taken in the preceding steps (i.e., the creation of the English description and flow chart and the writing of the program). Finally, the program must be *documented*. Docu-

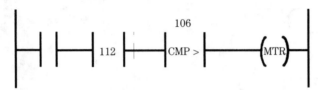

Figure 10.3 Ladder rung for the control of steam temperature.

mentation involves preparing a record of the program and the hardware connections, etc., so that subsequent users of the program will not experience difficulty.

The process presented above has been simplified so that the basic elements of programming are apparent. Many details specific to a given PC have been omitted. For example, a specific make and model PC might have certain registers set aside for I/O data and others set aside for general user storage (for information such as set points). These register numbers can be found by consulting the user's manual. The proper register must be addressed by the program, or the program will not perform as desired. Likewise, the input and output voltages that can be handled by different PC I/O terminals can vary. Care should be taken not to assign a 115-volt ac input voltage to a terminal capable of handling only 24 volts ac. Careful reading of the PC's user's manual is usually sufficient to prevent this type of mistake.

Applications

As noted in chapter 1, very few industries exist today that do not employ programmable controllers. Although most of these businesses are in the manufacturing sector, many in the service sector also use PCs. Industries as diverse as aerospace, automotive, bottling and canning, chemicals, entertainment, food and beverage, gas and petroleum, lumber, machining, metals, mining, packaging, petrochemicals, plastics, power, pulp and paper, rubber, and transportation all employ PCs in abundance.

This section provides several examples of how PCs are used in real industrial applications. Before we examine these examples, a number of standard symbols are introduced that are used in the figures that follow. These symbols are shown in Figure 10.4 and should be self-explanatory. An example of how these symbols are used is given in Figure 10.5, which is merely our previous example of steam temperature control. A comparison of Figures 10.5 and 7.2 indicates the usefulness of these symbols.

Process measurement

As the name indicates, programmable controllers find great utility in controlling various processes. Before a process can be controlled, however, key variables in the process must be measured. It is not always convenient to measure key process variables directly. One example is the mass flow rate of gas. Mass flow rate can be calculated if differential pressure, absolute pressure, and temperature are known. The relationship between mass flow rate and differential pressure, absolute pressure, and absolute temperature is given in equation 10-1:

$$\text{Mass Flow Rate} = K(\Delta P \times P_{absolute}/T_{absolute})^{1/2} \qquad (10\text{-}1)$$

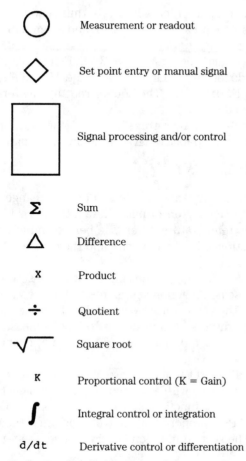

○   Measurement or readout

◇   Set point entry or manual signal

□   Signal processing and/or control

Σ   Sum

△   Difference

X   Product

÷   Quotient

√   Square root

K   Proportional control (K = Gain)

∫   Integral control or integration

d/dt   Derivative control or differentiation

Figure 10.4 Standard control symbols.

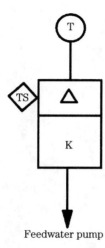

Figure 10.5 Control of steam temperature.

Feedwater pump

where K is a constant. The measurement of mass flow rate thus requires a differential pressure cell (for ΔP), an absolute pressure transducer, and a temperature transducer calibrated in degrees Kelvin.

A PC setup for the calculation of mass flow rate is shown in Figure 10.6. Input signals from the ΔP, P, and T transducers are fed into the PC. The ΔP and P signals are multiplied, and this product is divided by T. The square root of this term is then taken, and the whole term is scaled by the factor K.

Another process variable calculated is heat transfer rate. Heat transfer rate is a function of flow rate and change in temperature (ΔT), as shown in equation 10-2:

$$\text{Heat Transfer Rate} = C \times \text{Flow Rate} \times \Delta T \tag{10-2}$$

A PC setup for the calculation of heat transfer rate in a heat exchanger is shown in Figure 10.7. In the figure, the calculation of mass flow rate is provided by the setup shown in Figure 10.6. The temperature difference before and after the heat exchanger is taken and then multiplied by the flow rate and the constant C, yielding the heat transfer rate readout.

Another useful application of PCs in process measurement is the setup shown in Figure 10.8. This figure shows a sealed reaction vessel in which a chemical reaction is occurring. The temperature in the vessel must be controlled because the reaction becomes explosive at high temperatures. It is desirable to have a measurement of the reaction vessel temperature, the safety margin, and the rate of change of temperature in the vessel.

All of these measurement and control functions are shown in Figure 10.8. The temperature in the vessel, T, is measured with a thermocouple. This signal is used in performing all four functions. The first function is proportional control of temperature. The control signal from the PC's proportional control program is then supplied to the reactor cooling system. The second function is very simple: the temperature

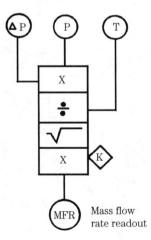

MFR Mass flow
     rate readout

Figure 10.6 Measurement of mass flow rate.

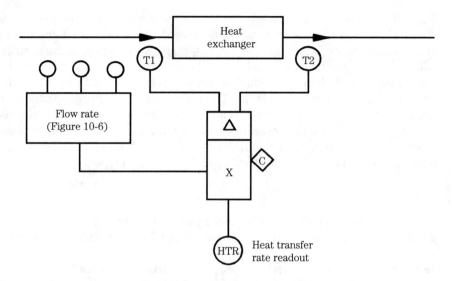

Figure 10.7 Measurement of heat transfer rate.

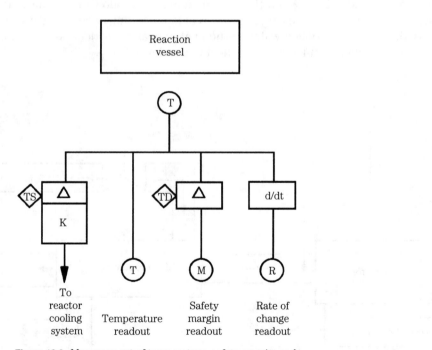

Figure 10.8 Measurement of temperature, safety margin, and
rate of change of temperature.

T is displayed on a control panel. The third function is the calculation and display of the safety margin. The safety margin is defined as the difference between reaction vessel temperature T and the danger temperature TD, which is the temperature at which the reaction becomes explosive. The final function is the calculation and display of the rate of change of temperature, which is simply the derivative of the temperature signal with respect to time. The portion of the PC program used to calculate the derivative term for PID control can be used for this purpose. An increase in the rate of change of reaction vessel temperature indicates that a safety problem might be developing. Generally, this indication will occur well before the safety margin or temperature readouts indicate that a problem has developed.

Process control

It is in the area of process control that PCs are most useful. The examples that follow represent just a few of the many ways that PCs are used in control applications.

Temperature control. Figure 10.8 presents a setup in which a PC is used to control temperature in a reaction vessel. Figure 10.9 shows a typical reactor and cooling system in greater detail. The chemical reactor is fed by a reactant stream at a specified temperature T. The temperature of the reactant stream is controlled by the valve in the cooling water line. When the valve is opened, the flow of cooling water to the heat exchanger is increased, and the reactant stream is cooled. As the valve is closed, the flow of cooling water to the heat exchanger is decreased, and the reactant stream temperature climbs to the ambient level.

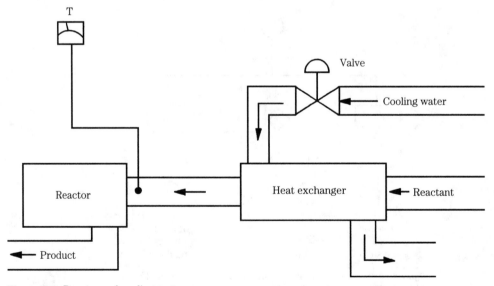

Figure 10.9 Reactor and cooling system.

In this uncontrolled (that is, not automatically controlled) system, a human operator would be assigned the responsibility of monitoring the temperature and adjusting the rate of flow of cooling water accordingly. In other words, the operator becomes the feedback mechanism. It is likely that the operator has other responsibilities and is not available for continuous measurement and control.

If it is assumed that the operator measures the temperature and adjusts the cooling water flow rate four times per hour, then Figure 10.10 could represent the temperature-versus-time plot for the system over an hour. The figure shows that, for much of the time, the temperature of the reactant stream is very different from the set-point temperature T. The reactor spends most of the hour either producing low yields of product (when the stream temperature is below the set-point temperature) or producing reasonable yields under unsafe conditions (when the stream temperature is above the set-point temperature).

A sudden upset or step change in reactant stream temperature is shown in Figure 10.11. If, as shown in the figure, the upset occurs shortly after the operator has adjusted the flow of cooling water, the result could be disastrous: the temperature could increase to the explosive limit before the next check is made.

An obvious improvement over this system is the use of a PC programmed with a proportional control algorithm, as shown in Figure 10.12. The difference between set-point temperature and stream temperature is taken, and the difference is multiplied by the gain K. This becomes the control signal used to adjust the electronic control valve in the cooling water line. The proportional control scheme in Figure 10.12 is shown in Figure 10.13 with standard control symbols.

Even with the offset inherent in proportional control (see chapter 7), the use of proportional control in this example greatly reduces the amount of time that the reactor spends in the unproductive (low-temperature) or unsafe (high-temperature) regimes.

Consistency control. Most refineries, chemical manufacturing plants, pharmaceutical plants, and paper mills are criss-crossed with pipes and filled with pumps. The most convenient method of moving raw materials or products back and forth in a plant is by pump. When the material to be moved is a liquid, usually no problems are associated with pumps. Generally, however, dry materials cannot be pumped.

If, however, the dry materials are first *slurried* (or mixed with water), they can be pumped, which is a common method of pumping pigments such as clay. Such a slurry is characterized by its *consistency*, which is just the percentage of dry material in the slurry. A pump that can handle the anticipated consistency is then selected.

Consistency control is important for two reasons. First, if the consistency becomes too thick, the pump will not be capable of moving the slurry. Second, the proper mixing of slurries for particular applications depends on the availability of a constant concentration of dry material in the slurry. Thus, consistency control is often a critical requirement.

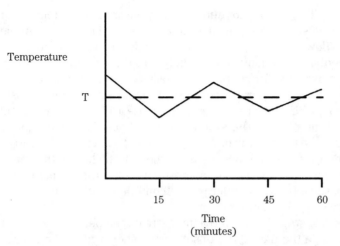

Figure 10.10 Temperature vs. time for an uncontrolled system.

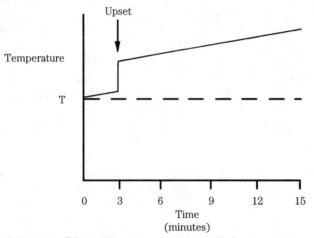

Figure 10.11 Effect of an upset on an uncontrolled system.

The first step in devising a consistency-control scheme is the selection of a consistency-measuring device. Such a device is shown in Figure 10.14. The slurry passes through a clear glass or quartz tube through which a light is transmitted. As the consistency of the slurry is increased, the light received by the detector is decreased (that is, the slurry becomes more opaque). When the consistency of the slurry is decreased, the light received by the detector is increased, and the slurry becomes more transparent. The detector signal is then treated by the electronics package within the measurement device and boosted to the appropriate level. The consistency mea-

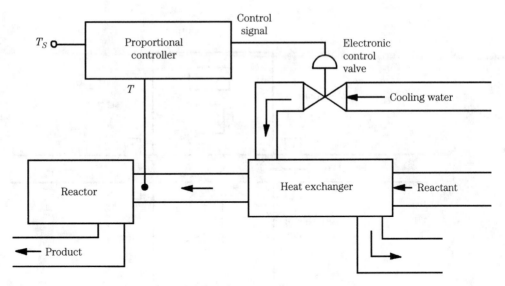

Figure 10.12 The proportional control scheme.

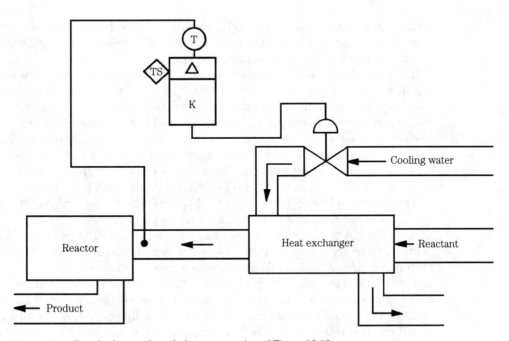

Figure 10.13 Standard control symbol representation of Figure 10.12.

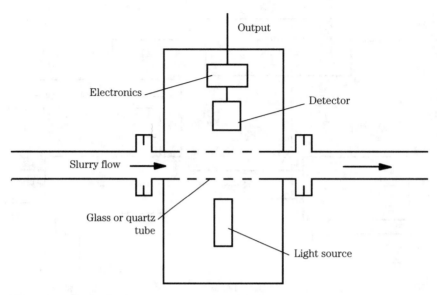

Figure 10.14 Device for the measurement of consistency.

suring device can be calibrated using slurries of known consistencies prepared in a laboratory. In general, the calibration curve would be different for different materials, such as clay, calcium carbonate, wood pulp, etc.

The control scheme for a simple, one-step dilution is shown in Figure 10.15 and is another example of proportional control. The consistency-measuring device is located on the outlet side of the pump because the pump is the only mixing element in this system, and the slurry should be well mixed before measuring consistency. The consistency signal is then compared with the set-point signal, and the difference is multiplied by the gain K. This control signal is then fed back to the electronic control valve in the dilution water line. If the consistency is too high, dilution water flow is increased. If the consistency is too low, dilution water flow is decreased.

The system in Figure 10.15 works best when small changes in consistency are required. If large dilution water flow rates are required, consistency control becomes erratic. Erratic consistency control can be remedied with the two-step dilution system shown in Figure 10.16. In the two-step system, the secondary dilution control step is the same as before. What is new is a primary dilution step in which the dilution water is mixed with the slurry in an agitated tank. (The agitated tank is a more efficient mixing element than the pump.) When the consistency error signal is large, such that the secondary control valve is almost fully opened, a second PC proportional control algorithm increases primary dilution water flow. Thus, the primary dilution system acts as gross adjustment to consistency, while the secondary dilution system fine-tunes the slurry to the desired consistency.

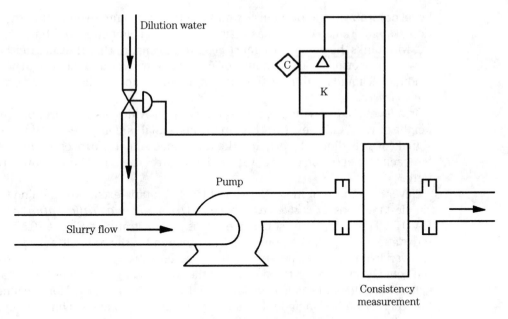

Figure 10.15 Proportional control scheme for a simple, one-step dilution.

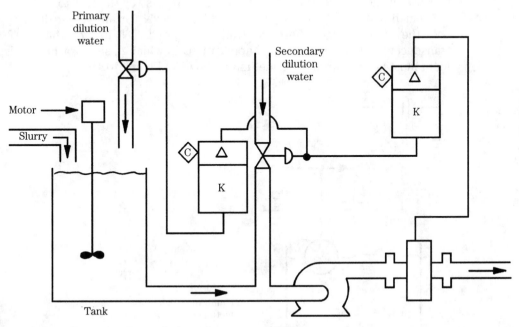

Figure 10.16 Proportional control scheme for a two-step dilution.

Level control. In the previous example, a tank was employed in the primary dilution system. Tanks are used as reservoirs, mixing elements, and buffers. In working with tanks, however, care must be taken to ensure that they are neither drawn down nor overfilled. The amount of mixture leaving a tank should just equal the amount of liquids entering the tank and can be accomplished very easily using *level control.*

In a level-control scheme, a level indicator provides a signal proportional to the height of the liquid level in the tank. This signal is compared with the set-point signal (desired liquid level), and the difference signal is processed by the PC and fed back to electronic control valves that control the input liquid flow rates to the tank.

Several methods of indicating liquid level in a tank exist. We examine two of the simplest methods: *variable-resistance* and *linear-voltage-differential transformer* (LVDT). The variable-resistance method is demonstrated in Figure 10.17. A vertical rod is attached to a float at one end. The other end of the rod is attached to the wiper of a variable resistor or potentiometer. A potential is applied across the resistor. As the liquid level changes, the position of the wiper changes, as does the voltage at the wiper. Thus, liquid level is transformed into a voltage reading. More accurate variations of this method include the use of the variable resistance in one arm of a Wheatstone bridge, as shown in Figure 10.18.

The principal drawback to this method is mechanical wear of the variable resistance. After a relatively short operating time, the variable resistor must be changed, because the abrasive action of the wiper tends to destroy the resistor.

A much better method of determining liquid level (or any linear displacement) uses the linear-voltage-differential transformer, or LVDT. In fact, the LVDT is the transducer of choice in almost all applications in which position or linear displacement must be measured. This method is shown in Figure 10.19.

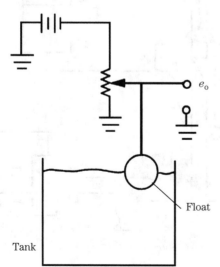

Float

Tank

Figure 10.17 The variable resistance method of indicating liquid level.

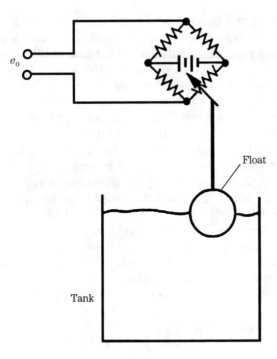

Figure 10.18 The variable resistance method of indicating liquid level, employing a Wheatstone bridge.

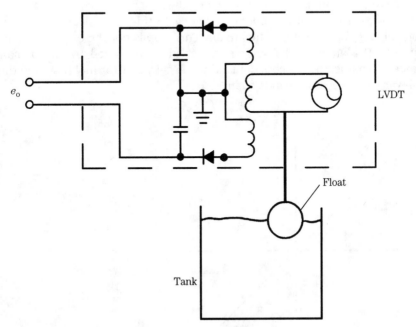

Figure 10.19 LVDT method of measuring liquid level.

A central coil excited by an ac signal is connected to the vertical float rod. As the level changes, the coil's position between the pair of coils is altered. If the moving coil is centered between the pair, equal voltages are induced in both. The rectified signals from the pair are of equal magnitudes, so that the output signal is zero. If, however, the moving coil is not centered between the pair, voltages of unequal magnitudes are induced in the pair of coils, and the output signal is nonzero (either positive or negative). This signal is used to indicate liquid level or linear displacement and is fed to the PC for processing. The output from an LVDT is not linear over all ranges, as is shown in Figure 10.20. With a little care in the choice of an LVDT, however, the output can be made linear over the range of interest.

The completed level-control scheme employing proportional-integral control is shown in Figure 10.21. In the figure, it is assumed that process demands control the outflow from the tank. In other words, the only degree of freedom in controlling tank level is derived from controlling the flow rate of liquid entering the tank. The LVDT provides a signal proportional to the liquid level, which is compared to the set-point signal. This output is multiplied by the gain K, integrated, and added to itself to provide the proportional-integral control signal. Because of the tank's capacity, oscillations are not a problem, so derivative action is unnecessary (see chapter 7).

Fuel-to-air ratio control.   The preceding examples dealt with measurement and control of processes involving liquids. This example covers new territory: combustion processes.

Almost all combustion devices, from the internal combustion engine of an automobile to the combustion furnace of an industrial power boiler, require an appropriate fuel-to-air ratio for maximum efficiency. In this example, we consider a process control loop for the ratio of fuel to air. This loop is not the only control loop for the furnace; the amount of fuel required is also controlled by the demands of the process. That is, the process demand is the set point for fuel flow, and the set point will vary as process demands change.

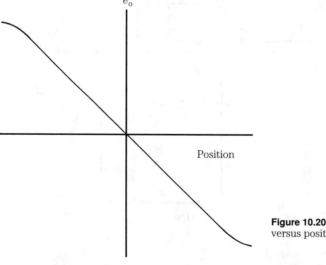

Figure 10.20 Output voltage versus position for the LVDT.

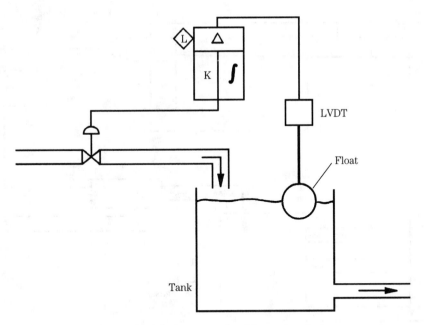

Figure 10.21 Level control employing proportional-integral control.

Figure 10.22 shows a furnace whose fuel flow rate is controlled by a proportional-integral control algorithm. It is assumed that this is a furnace generating steam to produce electrical power. As demand for electrical power increases (i.e., as the set point changes), the proportional-integral controller increases fuel flow to the furnace.

In this particular system, only the fuel flow to the furnace changes as demand changes. The air flow remains constant, so that the fuel-to-air ratio departs from the optimum value. The result is a mixture that is either fuel-rich, with incomplete combustion, wasted fuel, and increased pollution, or fuel-lean, with insufficient power generation.

This situation can be remedied with the control scheme shown in Figure 10.23. Both fuel and air flow rates are measured using the technique shown in Figure 10.6, and the ratio is obtained. This ratio is compared to the set-point value. The error signal is integrated and added to itself, and then used to feed an electronic control valve in the air flow line.

This example indicates some of the complexities of process control in industry. Many control loops are used to control one piece of equipment, and the effect of controlling one variable such as fuel flow on another such as fuel-to-air ratio must be considered.

Control of steam pressure. We now examine the control of steam pressure. Steam-pressure control is often critical not only for process efficiency but also for safety. If the pressure limit of a vessel is exceeded, it can explode. For this reason, oscillations in steam pressure are not desired, and proportional-integral-derivative control is needed.

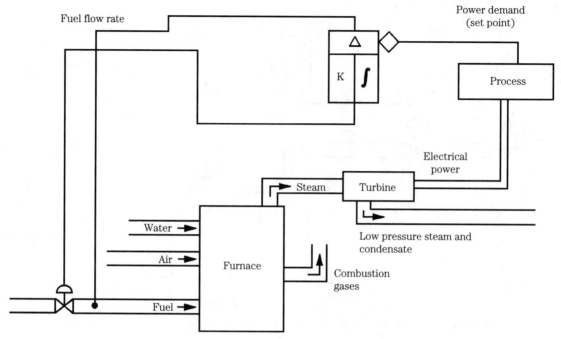

Figure 10.22 Power boiler proportional-integral control for fuel flow.

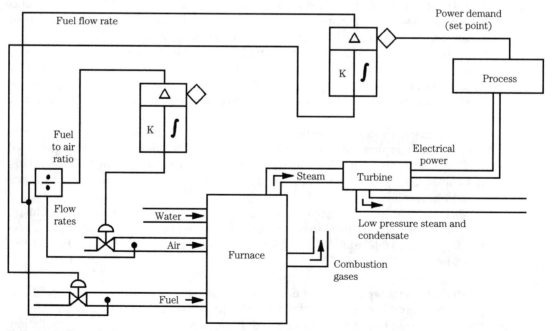

Figure 10.23 Power boiler with proportional-integral control for fuel flow and proportional-integral control for fuel-to-air ratio control.

A simple steam-pressure control scheme is shown in Figure 10.24. In this figure, all the considerations of fuel flow, fuel-to-air ratio, and ignition signal are represented schematically by the firing rate controller block to simplify the figure. A pressure transducer in the steam line indicates the steam pressure, and the signal from this transducer is compared to the set-point value. The error signal is then subjected to integration and differentiation, and the combined signal (the PID control signal) is then fed to the firing rate controller.

An added complication occurs when process demand is considered, as illustrated in Figure 10.25. When the demand for steam changes rapidly, steam pressure tends to fluctuate. In the system shown in the figure, total steam demand controls the fuel flow rate. (Fuel-to-air ratio is controlled in the firing rate controller block.) The steam-pressure feedback loop acts as a trimming control. Multiplication of the two PID signals is appropriate because the amount of correction necessary for a given deviation in steam pressure is proportional to process steam demand (or load). By multiplying the signals, one control signal compensates equally for both process demand and pressure variations.

Most plants use more than one steam or power boiler, which rapidly complicates the control scheme. The general principles outlined above, however, are still used in controlling steam pressure and flow.

Effluent discharge control. Many process industries use large quantities of water, which is why so many industries are located near lakes or rivers. After the water used in manufacturing is recovered, it is purified and discharged back into the river or lake. Strict regulations cover the discharge of process water. The biological oxygen demand (BOD) of the discharged water is limited to certain ranges, the color of

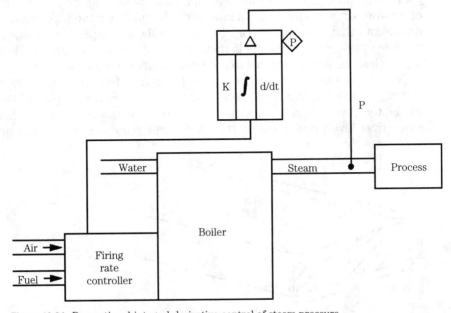

Figure 10.24 Proportional-integral-derivative control of steam pressure.

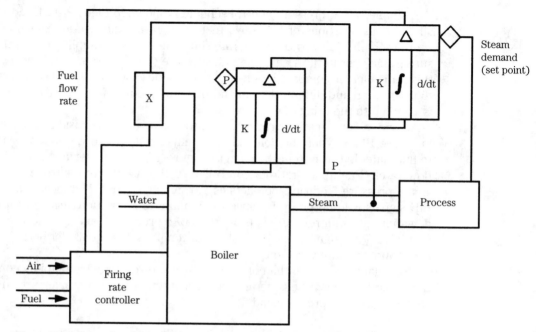

Figure 10.25 Proportional-integral-derivative control of steam pressure and steam demand.

the discharged water is monitored, and the effect of discharged water on river or lake pH and temperature are strictly controlled.

In this example it is assumed that all the above effects can be controlled by adjusting the water discharge rate to an optimum value compared to river flow rate. In other words, it is assumed that a ratio of discharged water volume to river water volume exists that is optimum for control of BOD, color, pH, and temperature. The essence of the control scheme is then control of effluent volumetric flow rate. The feedback signal is proportional to the river volumetric flow rate.

Three inputs are needed to determine the volumetric flow rate of the river: river velocity, river level, and river width, shown schematically in Figure 10.26. If it is assumed that the river has a smooth, level bottom of known width, the height of the river (river level) multiplied by the width gives the cross-sectional river area (the

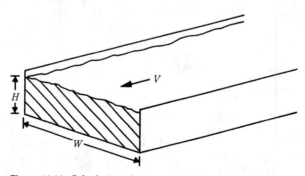

Figure 10.26 Calculation of river volumetric flow rate.

shaded portion in Figure 10.26). Multiplication of this area by river velocity gives the volume of water flowing past the shaded area for a given time. For example, assuming that the river level is 3 feet, the river width is 10 feet, and the river velocity is 60 feet-per-minute, the volumetric flow rate is

$$Q = 3 \text{ ft} \times 10 \text{ ft} \times 60 \text{ ft/min}$$
$$= 1800 \text{ cubic feet per minute} \tag{10-3}$$

Using the conversion factor of 7.48 liquid gallons per cubic foot yields

$$Q = 1800 \times 7.48$$
$$= 13,464 \text{ gallons per minute} \tag{10-4}$$

In general, river level and river velocity are not completely independent. That is, a higher river velocity is most likely found when river level is higher than normal. To assure proper calculation of volumetric flow rate, both river level and river velocity should be measured.

In the effluent control system presented, a pump is used to move the effluent from the reservoir to the river. The gallonage moved by the pump can be controlled continuously from zero gallons per minute to the pump maximum value by the pump speed control unit. The control scheme is shown in Figure 10.27. Electromagnetic-type flow meters are used to measure river velocity and effluent velocity. Effluent velocity is multiplied by pipe cross-sectional area (previously measured) to obtain effluent volumetric flow rate.

The river velocity signal (from the electromagnetic flow meter) is multiplied by the river level, and this product is multiplied by river width, which was previously measured. This signal, the river volumetric flow rate, is divided into the effluent volumetric flow rate, yielding the ratio of effluent to river volumetric flow rates.

The ratio of effluent volumetric flow rate to river volumetric flow rate is now compared to the optimum value (set-point value). The error signal is subjected to PID treatment, and the control signal is fed to the speed control box for the pump. Deviations from the ideal ratio are corrected by adjusting the effluent flow rate.

Other examples

The examples presented previously concentrated primarily on PC control of unit processes typical of those found in the chemical industry. PCs are also valuable in controlling other process steps, and an example is shown in Figure 10.28. This figure illustrates the use of a PC to control the processing of olives and was taken from supplier literature for the GE Fanuc Automation line of programmable controllers.[*] Other PC applications include high speed label lines (Figure 10.29) and control of wine presses (Figure 10.30).

---

[*] Note that in this and the following GE Fanuc figures, GE Fanuc Automation makes no representation or warranty, expressed, implied, or statutory with respect to, and assumes no responsibility for the accuracy, completeness, or usefulness of the information contained in these figures. No warranties of merchantability of fitness for purpose shall apply.

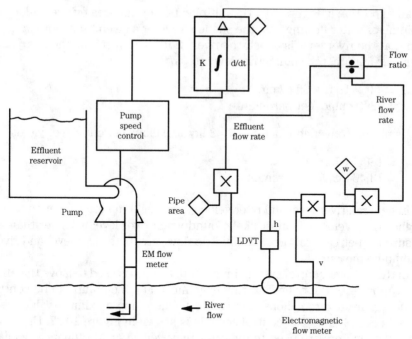

Figure 10.27 Effluent discharge control scheme.

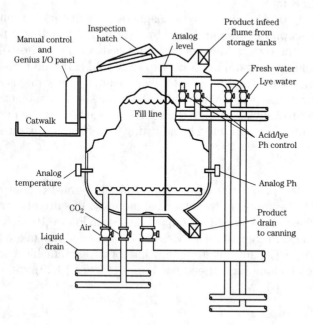

Figure 10.28 PC control of olive processing (reprinted with permission of GE Fanuc Automation North America, Inc.).

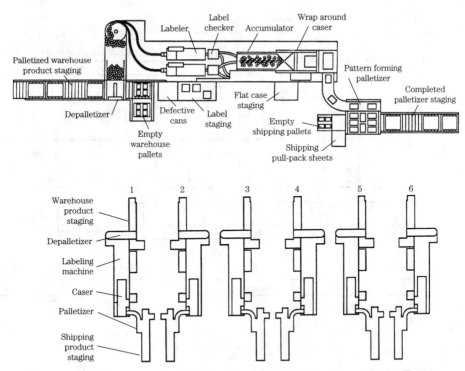

Figure 10.29 Control of a high-speed label line (reprinted with permission of GE Fanuc Automation North America, Inc.).

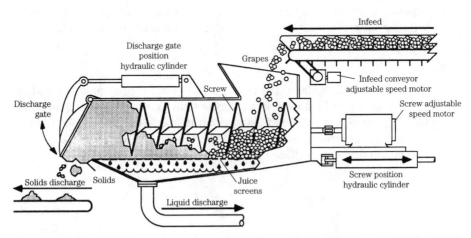

Figure 10.30 Control of wine pressees (reprinted with permission of GE Fanuc Automation North America, Inc.).

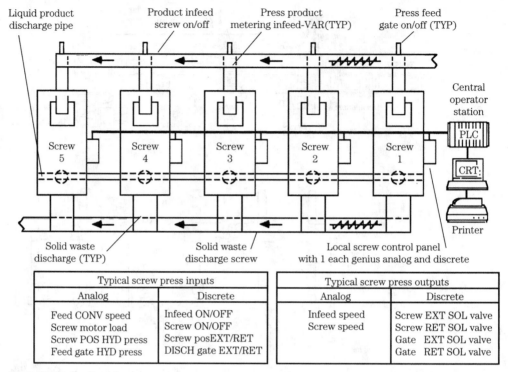

Figure 10.30 Continued.

Catalogs from suppliers of programmable controllers are very useful in providing typical examples of PC usage. You should consult the supplier literature or contact the supplier's representative for other potential applications of programmable controllers.

Summary

This chapter has outlined the basic elements of PC programming and described several examples of PC usage in the process industries. These examples should make clear the great advantage of using programmable controllers for the control of various industrial processes. Additional insights on the use of PCs in process industries can be obtained from PC supplier literature.

Local Area Networks

The subject of communications between programmable controllers (PCs) and peripheral devices was introduced in chapter 8. That chapter specifically dealt with the transmission of data and control instructions between a PC and a peripheral. The focus of such a communications system is the complete transmission of data and instructions within one scan of the PC. Such a network of PCs and peripherals is often called a *control network*.

This chapter also addresses communications, but the emphasis is slightly different than that of chapter 8. This chapter emphasizes communications between many PCs, or between PCs and intelligent devices such as computers, intelligent cathode ray tubes (CRTs), or special I/O interfaces. In this case, the focus is on data sharing and system status sharing, not the communication of I/O data within the time frame of a single scan. This type of network is often called a *communications network*.

In principle, communication between two PCs (or between a PC and an intelligent device) is easy to provide. One simply connects the input of the first PC with the output of the second PC, and the output of the first with the input of the second, using pairs of wires. This is, in fact, the method originally used to provide communications between PCs.

This method worked very well. The rate of data transmission was limited only by the baud rates of the two PCs because only one bit of data was carried by a pair of wires. The limit of one bit per pair of wires did require that a larger quantity of wire be used, but the cost was not prohibitive when just two PCs (in relatively close proximity) were connected. For two PCs physically separated by a large distance, however, wiring costs became significant.

With the addition of a third PC, communications became a more complicated affair. Using pairs of wire to interconnect all the PCs was prohibited because of the increased wiring costs. A central computer was then employed to handle the communications between PCs. In the central computer scheme, three or more PCs are connected to a central computer in a configuration similar to the star configuration of chapter 8. All communications between PCs are thus routed through the central computer. Although this system minimizes wiring costs, it limits transmission rates to the baud rate of the central computer. Also, if the central computer fails, the entire system goes down.

Given the limitations of the two previous methods of communicating between PCs (wire pairs and central computer), it is understandable why the *local area network* was developed.

A local area network is defined as a system that allows the transmission of data and system statuses between PCs (or between PCs and intelligent devices) at high speeds over long distances (i.e., two to three miles of cable). A local area network, sometimes known as a *data highway*, can support many PCs or intelligent devices—most local area networks can support at least 100 different PCs or nodes. (Any PC or intelligent device connected to a local area network is called a *node*.) Typically, the communications rate of a local area network is on the order of 57 or 58 kilobaud or greater.

Local area networks for PCs are quite similar to the local area networks of business systems, which are often used to connect office equipment such as word processors to a central computer and also allow communications between word processors and computer terminals scattered in various office locations. The two primary differences between business systems local area networks and PC local area networks are that local area networks for business systems are slower than PC local area networks, and that business systems networks are not shielded from electromagnetic noise and interference. With some work, business systems local area networks can be adapted for use with PCs.

In addition to sharing data and system status, local area networks can be used in *distributed control*. In distributed control, the control functions of one large PC are distributed among several PC subsystems. Currently, distributed control is a very popular and important control concept. An in-depth study of distributed control is beyond the scope of this book, however, and readers desiring additional information are referred to the references in the bibliography.

There are two major local area network formats. The first format includes those networks controlled by a master computer or PC. Communications between PCs in the network are routed through the master computer or PC. This format is known as a *master-slave* local area network (Figure 11.1). Although this system allows for total control of communications in the network through the use of a master computer, it has the same disadvantage as the central computer scheme discussed previously; when the master computer (or PC) fails, the entire network goes down. Master-slave networks generally include a backup master computer or PC, which is designed to operate the network when the master computer fails.

The second format is one in which there is no master to control the network. Each device, or node, in the network controls its own communications. This format is known as the *peer-to-peer* local area network (Figure 11.2). Control of such a network is rotated among the various nodes in the network. In other words, PC 1 has temporary control of the network to effect communications with other network PCs. At some later time, PC 1 relinquishes control of the network, and PC 2 takes control.

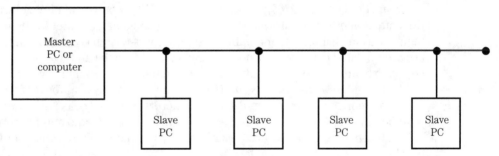

Figure 11.1 Master-slave format local area network.

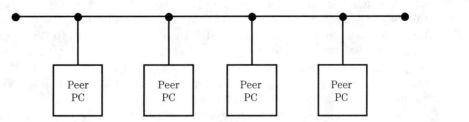

Figure 11.2 Peer-to-peer format local area network.

After PC 2 completes its communications, it passes control of the network to PC 3, and so on. The passing of network control from one peer device to another is known as *token passing*. That is, the PC that has the token has control of the network. The PC in control loses that control when it passes the token to another PC. The peer-to-peer local area network has the advantage that failure of one PC does not bring the system down. On the other hand, the disadvantage is that the schemes necessary to guarantee satisfactory network control are quite complicated, as each peer PC shares control responsibilities equally with other network PCs, which is discussed later in this chapter.

Transmission Media

The term *transmission media* refers to the physical components that allow communications between PCs or between PCs and intelligent devices. The two major components of transmission media are the modem, or network adapter module, and the communications medium, or bus.

The term *modem* is an acronym for MOdulator-DEModulator; however, the term *network-adapter module* is more descriptive of its function. The modem serves as an interface between the PC or intelligent device and the communications medium (wire, cable, etc.). Essentially, the modem converts the two-level binary voltage from the PC into a two-frequency audio signal for transmission over the wire or cable. Likewise, when the modem receives an audio signal, it converts that signal to a binary voltage for use by the PC. The modem uses a technique known as *frequency-shift keying* (FSK) to accomplish its modulation-demodulation task. Usually, each PC or node in the network has its own modem. Some modems, however, can accommodate several network devices.

The modem is capable of receiving data from a PC at one baud rate and transmitting the data at another, usually higher, baud rate by including a buffer in the modem (remember from chapter 8 that a buffer is simply short-term memory). Data from the PC are accumulated in the buffer and then transmitted at high baud rates over the bus. Received data are stored in the buffer and transmitted to the PC at a lower baud rate.

The term *communications medium* refers to the physical form of the bus. The most common media are twisted-pair conductors, coaxial (coax) and triaxial cables, telephone lines, fiberoptics, and electromagnetic waves (radio and microwave).

Twisted-pair conductors are described by their name. They consist of two single conductors twisted about each other. Until relatively recently, the twisted pair was the most widely used communications medium. It has now been outdistanced by the cables, which are discussed next. The principal advantage of the twisted-pair conductor is its low cost. When shielded, the twisted-pair conductor is relatively immune to noise and interference, and transmission rates up to 250 kilobaud are attainable. The principal disadvantage of the twisted-pair conductor is its nonuniformity. This nonuniformity is manifested in a varying characteristic impedance, which makes impedance-matching virtually impossible, resulting in a limited range for the twisted-pair conductor (4000 feet or less).

Cable (coaxial and triaxial) represents the most widely used communications medium. Its high degree of uniformity (uniform characteristic impedance) allows its use over large distances of up to two to three miles. The use of repeaters (devices that receive a signal and automatically retransmit, or repeat, the signal in amplified form) in a network allows the use of cable over distances of 30 miles or more.

Two basic types of coaxial (coax) cable exist: *baseband* and *broadband*. Baseband (or passive) coax allows the transmission of only one signal at a time. Broadband (or cable TV) coax allows the transmission of two or more signals simultaneously on different channels. One broadband channel really consists of two frequencies: a high frequency for transmission and a low frequency for reception. The transmission rate on a broadband channel is usually higher than that for baseband coax (10 megabaud versus 2 megabaud), but the cost of broadband coax can be twice as great as the cost of baseband coax. Broadband coax is generally used when the local area network forms a part of a broadband network. In a broadband network, one channel is reserved for PC communication, while the remaining channels would carry other signals, such as video, computer terminals, monitors, etc.

The use of telephone lines as a communications medium is not an unfamiliar concept to users of computers. A modem designed for use with a telephone handset can be used to provide communications between personal computers all over the country or between intelligent terminals and mainframe computers. The same is true for PC networks. With modems and telephone lines, the PC "local" area network can encompass a very large locality. The cost, of course, is in the form of long-distance telephone bills.

Fiberoptics as a communications medium is in its infancy and is expected to grow rapidly as technological developments occur. Many reasons exist for optimism regarding fiberoptics. They are completely immune from electromagnetic noise and interference, smaller and lighter than coax cables, and capable of supporting amazingly high transmission rates (several hundred megabaud) over long distances (five to six miles). The primary losses associated with fiberoptics occur when the bus is tapped (i.e., when a connection is made on the bus to attach a device). Progress in fiberoptics technology is occurring rapidly; one long distance telephone company is, at the time of this writing, in the process of installing lines comprising 100 percent fiberoptics technology. Widespread use of fiberoptics in PC local area networks should not be far behind.

Electromagnetic waves (radio and microwave) find limited use as a communications medium in PC local area networks. Offshore operations, such as drilling rigs, are the primary users of this medium at present. In theory, the use of microwaves and satellites in geosynchronous earth orbit would allow communications between PCs located a continent apart.

Topologies

The word *topology* is used to describe the physical arrangement or configuration of nodes in a network. The daisy chain and star configurations shown in Figure 8.6 are topologies for multiple peripherals. The star topology shown in Figure 11.3 is similar to the star topology shown in Figure 8.6. The only difference is that Figure 11.3 shows a star topology involving PCs, not peripherals, with a master PC or computer in the center of the star. Obviously, the star topology works only with the master-slave local area network format. The advantage of the star topology is that communications between each slave and the master can occur whenever needed. The primary disadvantages are that transmission rates are limited by the baud rate of the master, through which all communications must occur, and that failure of the master brings the entire system down. Compared with some of the other topologies, wiring costs can also be high.

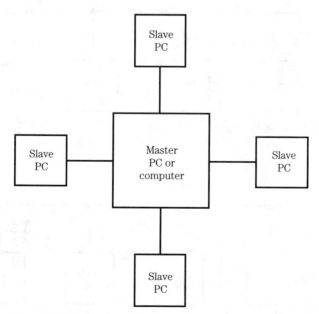

Figure 11.3 The star topology.

The daisy chain topology, shown in Figure 11.4, is used in peer-to-peer format local area networks. A more popular topology, however, is the *common bus* topology, shown in Figures 11.5 and 11.6. Figure 11.5 shows the common bus topology in a peer-to-peer format, while Figure 11.6 shows the common bus topology in a master-slave format.

The peer-to-peer common bus topology in Figure 11.5 is sometimes called a *multidrop* topology. The bus itself is a cable tapped at each peer PC. The tap used is called a *tee tap*. A short length of cable (drop length of 500 feet or less) is used to connect the main cable (bus) to the modem at each peer PC. The cable that connects the peer PC modem to the bus is called a *drop cable*. When Figure 11.5 is compared to Figure 11.4, it becomes apparent that the daisy chain topology is simply a peer-to-peer common bus topology with a zero drop length.

As with other peer-to-peer formats, this common bus topology allows PCs to communicate with each other directly; information is not passed through a master PC. Because no master is present, a method of controlling network transactions is necessary that allows the PCs to transmit and receive data without interfering with the transmission or reception of another network peer. Methods of controlling transactions on this type of network are covered later in this chapter.

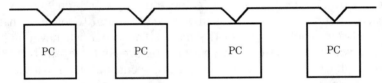

Figure 11.4 The daisy chain topology.

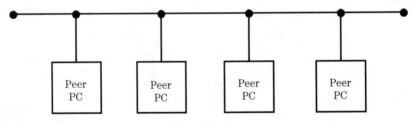

Figure 11.5 The common bus topology: peer-to-peer format.

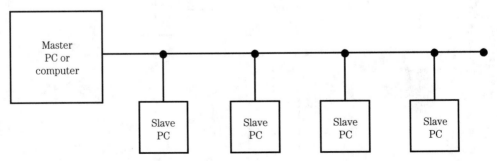

Figure 11.6 The common bus topology: master-slave format.

The principal advantage of this type of topology is minimal wiring costs. The principal disadvantage is the dependence on a single bus for communication between all PCs in the network. Because a single bus is used, communications between PCs are not as rapid as in other topologies. Also, failure of the bus brings down the network.

The common bus topology shown in Figure 11.6 is a master-slave format local area network. In this topology, all communications are controlled by the master PC or computer. The master communicates with the slave PCs: any communication between slaves is routed through the master. The master PC addresses, or *polls*, each slave PC for data. When polled, the slave transmits data to the master. The master then has the responsibility of distributing this information to other slaves. This topology re-emphasizes the disadvantage of other master-slave local area networks: failure of the master brings the system down.

Figure 11.7 shows the *loop* topology. The loop topology is not often used in industrial applications, primarily because the failure of any one PC in the loop is sufficient to bring the entire network down.

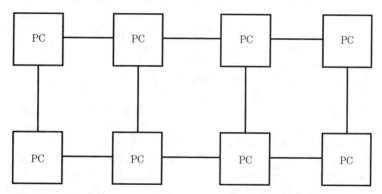

Figure 11.7 The loop topology.

Network Transactions

The two principal methods of controlling network transactions, polling and token passing, have been mentioned previously. Polling is used in master-slave format local area networks, while token passing is used in peer-to-peer format local area networks. A third method, *collision detection*, is available but not often used in local area networks.

The principle behind the polling method of accessing the local area network is simple. The master PC or computer asks (polls) a specific slave to transmit data. The master PC or computer then waits a specified period of time. If no data are transmitted by the slave, the master moves on to the next slave in the network and assumes that the nonresponsive slave is dead (nonfunctional). The master continues to poll each slave in the network. Communication between slaves in the network is slower than in peer-to-peer format networks because communication between slaves requires that each slave first be polled and that the master act as an intermediary in sending the message. This type of polling is often called *query-response* communication.

The polling of each individual slave PC in the query-response communication is time consuming. The length of time necessary to communicate with the slaves can be shortened by using a variation on this technique. This variation is known as the *broadcast* technique. In a broadcast transaction, the master sends the same message to all the slaves simultaneously. Each slave PC receives the message at the same time and carries out the orders or instructions that have been broadcast, reducing the length of time necessary for transmitting instructions. Responses from individual slave PCs, however, still require the polling of slave PCs by the master.

Communication between peer PCs in a peer-to-peer local area network requires a system whereby control of the network is passed from peer to peer. As described previously, token passing is the preferred technique for this purpose. In some respects, token passing is similar to polling. The difference is that each peer in the system takes a turn acting as master of the system. The peer PC that "possesses" the token (i.e., has control of the network) acts as master and polls the other PCs. This technique can be thought of as *distributed polling*. The PC with the token passes the token to the next PC upon completion of its communications. If the next PC does not accept the token, the first PC assumes that particular node is dead and attempts to pass the token to the next PC in the network. This process is similar to passing the baton in a relay race: the runner with the baton is the only one allowed to run, and the next runner cannot begin until he receives the baton. Token passing can be time consuming because each PC is given control of the network for a specified period of time, often called the *token-holding time*.

The PC in control of the network (i.e., the PC with the token) can communicate with another PC in the network by using a *handshaking signal*. The handshaking signal actually consists of two distinct signals. The PC sending the message first sends a request-to-transmit signal to the target PC. The target PC, if capable of accepting the message, then sends a clear-to-transmit signal back to the original PC. The transmission of both signals constitutes a handshake. When the handshake is completed, transmission of data between the two PCs is allowed to proceed. The nature of the handshaking signal varies with the communications protocol used by the network.

Collision detection is a form of network transaction more popular in business systems networks than in local area networks. Collision detection works as follows: each peer PC in the network is equipped with signal-sensing circuitry. This circuitry detects transmissions on the bus and prevents a PC from transmitting while another PC transmits. When two PCs begin transmitting at the same time, however, a collision of data on the bus is detected. The signal detection circuitry in each PC then disables the PC so that transmission of data is interrupted. The PCs then wait a specified period of time and attempt to retransmit the data. In other words, no token passing occurs in this system. A PC can gain control of the network whenever no other PC is transmitting. Simultaneous transmission by two or more PCs disables the network for a short period of time.

This system works well, provided there are few nodes in the network. Unfortunately, large networks cannot use collision detection because the speed with which data are transmitted is reduced each time a collision is detected, and the probability of collision increases with the number of nodes. Thus, collision detection systems are more popular with the (usually smaller) business systems networks.

It should be apparent from this summary of network transactions that the software, or programming, necessary to run the network is fairly complicated. Most PC network suppliers provide software tailor-made for their local area networks. The compatibility of a particular make and model PC with the various local area networks should be a major consideration in the selection of a PC system.

Protocol

In chapter 8, communications standards commonly used for communication between PCs and peripherals were reviewed. These communications standards are often called *protocols*. A protocol is a standard set of rules that specifies the format or mode of communication between devices. In this case, a device means the same as it did in chapter 8: a PC, printer, or other peripheral device. The protocol, or communications standard, also includes mechanical aspects such as wiring, as well as signal levels and timing.

In chapter 8, four protocols were mentioned: RS-232C, RS-449, 20-mA current loop, and IEEE 488. In addition to these four protocols, two other protocols work well for local area networks. These are the ISO Open Systems Interconnection Reference Model and the IEEE 802 standard. The physical details of these standard interfaces can be obtained from the International Standards Organization (ISO) and the Institute of Electrical and Electronics Engineers (IEEE), respectively.

The ASCII code (see chapter 8) remains the most popular protocol for the translation, or encoding, of alphanumeric characters into binary. As mentioned previously, the ASCII character code is given in appendix D.

One protocol that has gained immense popularity in recent years is the Manufacturing Automation Protocol (MAP). This protocol, introduced by General Motors, allows communications between various devices (PCs, peripherals, etc.). Both hardware and software have been developed for compatibility with MAP, seen in chapter 12. MAP is supported by the ISO and IEEE.

Summary

The local area network provides a convenient method by which communications between PCs are effected. The local area network allows system statuses and data to be displayed on many terminals at one time and allows for communications between PCs involved in distributed control. This chapter has provided a brief outline of the basic elements of local area networks. Supplier's literature should be consulted for more details.

12

Overview of Available Programmable Controllers

This chapter presents technical information on programmable controllers (PCs) available from domestic (U.S.) suppliers at the time of this writing. A list of domestic PC suppliers, with addresses and telephone numbers, can be found in appendix A. Further technical information on the PCs listed in this chapter can be obtained directly from the supplier.

Not all the PCs discussed in this chapter are manufactured in the United States. Some are manufactured abroad and marketed in the U.S., some are joint ventures between U.S. and foreign manufacturers, and some are manufactured solely in the United States. The companies listed in this chapter and in appendix A are thus designated as *domestic suppliers*.

It should be emphasized that the programmable controller market in this country changes very quickly. New PC suppliers spring up each year, and some are eliminated. And, of course, many of the older manufacturers remain competitive in the PC market year after year. Given the changes that occur, it is necessary to remind you that the information presented in this chapter is correct only for the time of this writing. The reader interested in staying up to date with current PC manufacturers and systems is advised to subscribe to two trade magazines: *Control Engineering* and *I&CS*.

ABB Process Automation, Inc.

This firm, a division of Asea Brown Boveri, Inc., markets the MasterPiece series of programmable controllers. The series includes two PCs: the MasterPiece 51 and the MasterPiece 90.

The MasterPiece 51 is a small PC featuring 4K of random access memory (RAM) with a battery backup system (see chapter 6). The I/O capability is a total of 32 points, all of which are discrete (ON or OFF; see chapter 7). It can handle either dc or ac signals. The programming language is a high-level, functional flow chart programming language (GRAFCET), and the PC can be programmed with either a cathode ray tube (CRT) or computer (see chapter 8). It cannot be programmed in ladder language.

The MasterPiece 51 is suitable for use in a local area network, in a master-slave format with a host computer (see chapter 11). It is capable of performing diagnostic (self-checking) functions.

The MasterPiece 90 is a larger PC featuring 128K of programmable read-only memory (PROM) and 256K of RAM with a battery backup system. It has a total of 256 I/O points, which include discrete and analog, ac and dc. In addition, the following special interfaces are available: proportional-integral-derivative (PID), ASCII, high-speed counter, and positioning.

The programming language is a high-level, functional flow chart programming language (GRAFCET), and the PC can be programmed with either a CRT or computer. It cannot be programmed in ladder language.

The MasterPiece 90 is suitable for use in a local area network, in either a master-slave format with a host computer or in a peer-to-peer format. It is capable of performing diagnostic functions and mathematical functions, such as arithmetic.

Active Systems Group, Inc.

Active Systems Group, Inc., of Redmond, Washington, markets the StepLadder™ programmable-logic controllers (PLC), which comes in two versions: the original StepLadder™ PLC, and the StepLadder™ PLC, version 2.1. Both units have 512 bytes of electrically erasable PROM (EEPROM) memory (see chapter 6), and a total of 24 discrete I/O points. I/O capabilities include ac, dc, and ASCII. In fact, the StepLadder™ PLC operates with just about any I/O module that uses a 50-pin header-type connector and a 5-volt power supply. (The 5-volt power supply for the I/O module serves as the power supply for the StepLadder™ PLC.) Two of these PCs can be connected together for a total of 48 I/O points.

The central processing unit (CPU) of the StepLadder™ PLC offers 24 internal relays (i.e., 24 relay ladder operations in a ladder program), two timers (0.01-second time base), and 10 counters that can be used as timers. In addition, it offers latching relays and one-shots (see chapter 9). The StepLadder™ PLC version 2.1 has 88 internal relays. Both use serial interfaces (see chapter 8).

These devices are programmed with STEPS™, a ladder language that was introduced previously (see chapter 9 and appendix C). They are programmed with a computer via an on-board serial interface. Both versions of this PC are suitable for use in networks in either master-slave or peer-to-peer formats. They can be used with remote I/O and offer continuous internal diagnostics.

Adatek, Inc.

Adatek, Inc., of Sandpoint, Idaho, has been mentioned previously (see chapter 9 and appendix B). Adatek produces the System 10 Programmable Controller, the ESE-2000 Industrial Controller, and the SE-2000 Single Board Controller for the IBM AT personal computer. Perhaps the most exciting contribution by Adatek, Inc., to the PC industry is its State Language approach to programming PCs, discussed in appendix B.

The System 10 Programmable Controller features 32K of user memory (complementary metal-oxide semiconductor, or CMOS RAM, and, erasable PROM, or EPROM). The RAM has a lithium battery backup system. The control unit (the CU10E) supports up to 24 digital I/O points, which can be expanded with the digital expansion frame (DX10) to a maximum of 1176 digital I/O points with any mix of ac or dc input or output signals. Additionally, the CU10E control unit can support up to 12 analog I/O frames (AN10) for a maximum of 96 analog I/O points (eight I/O points per AN10 frame).

Features of the System 10 include up to 3168 timers (depending on the number of additional I/O frames added), eight PID control loops, counters, one-shots, ASCII, and binary communications. It performs advanced mathematical calculations and diagnostics, and its minimum scan time is 1 millisecond. It can be used in master-slave or peer-to-peer networks.

The System 10 has two serial interface ports: one RS-232 and one that can be configured as either RS-232 or RS-422. Baud rates range from 300 to 9600. The System 10 can be programmed from any IBM compatible personal computer with an RS-232 port or any ASCII terminal (CRT).

The programming language, State Language, is discussed in some detail in appendix B. It offers programming from a plain-English-language description of the process to be controlled. In essence, the control system desired is described in plain English, and programming is virtually completed. This technique is among the highest levels of programming languages available today. Adatek, Inc., has developed this language broadly so that it can be used with other brands of PCs and I/Os (e.g., Opto 22, GE Fanuc Automation, and Allen-Bradley 1771).

The ESE-2000 Industrial Controller and the SE-2000 Single Board Controller offer similar features. Both have 512K of nonvolatile RAM (NOVRAM), and a maximum of 3968 I/O points (digital and analog, ac and dc signals). Both allow for a variety of I/O modules (high-speed counter, positioning control, and PID), and both employ the State Language approach to programming. Like the System 10, both are suitable for use in master-slave and peer-to-peer format networks, and both offer advanced math features and diagnostics.

Allen-Bradley Company

Allen-Bradley offers a wide variety of programmable controllers. These PCs fit into one of two lines: the SLC line or the PLC line. The SLC line of PCs is low-end (small, inexpensive, and suitable for simple control functions), and includes the SLC 100, SLC 150, SLC 500, SLC 5/01, and SLC 5/02. All these programmable controllers have CMOS RAM with battery backup. In addition, the SLC 100 and SLC 150 offer optional EEPROM. Memory sizes are small. The SLC 100 can hold 885 words in memory, while the SLC 150 can hold 1200 words. The SLC 500 holds 1K, and the SLC5/01 and 5/02 offer choices: 1K or 4K, and 4K or 16K, respectively.

The I/O capabilities of the SLC programmable controllers are as follows: SLC 100, 24 (analog, ac, dc); SLC 150, 72 (analog, ac, dc, high-speed counter, positioning); SLC 500, 256 (analog, ac, dc, positioning); SLC 5/01, 256 and SLC 5/02, 480 (both analog, ac, dc, high-speed counter, positioning, and PID).

All SLC programmable controllers can be programmed with ladder language. None can be programmed with a high-level language. Programming is accomplished manually or with a computer. They are suitable for use in networks, with remote I/O, and with a host computer. The entire SLC line (excluding the SLC 100) is capable of performing mathematical functions, and all perform basic diagnostic functions.

The PLC line of PCs includes the PLC-3 and PLC-5 series. The PLC-3 series consists of the PLC-3 and the PLC-3/10. The PLC-3 has 2 megabytes (M) of battery-backed RAM ($1M = 1K \times 1K = 1024 \times 1024 = 1,048,567$), while the PLC-3/10 has 128K of battery-backed RAM. The PLC-3 and PLC-3/10 support the same type of I/O modules (analog, ac, dc, high-speed counter, and positioning), but with different total I/O points (8192 for the PLC-3, 4096 for the PLC-3/10). Both are capable of ASCII communications and PID control. Both can be programmed with ladder or high-level languages via CRT, cassette tape, or computer.

The local area network capabilities of these two PCs include remote I/O, master-slave, and peer-to-peer formats. The Allen-Bradley data highway (or local area network) adheres to the manufacturing automation (MAP) protocol. Both PCs perform basic mathematical functions and diagnostics.

The PLC-5 series of programmable controllers includes the PLC-5/12, 5/15, 5/25, 5/40, 5/60, 5/250, 5/VME, and the PLC-5/30 programmable controllers. All but the PLC-5/VME and the PLC-5/30 offer battery-backed CMOS RAM with optional EEPROM. For the two PCs mentioned, the EEPROM is not optional. Memory sizes vary as follows: 5/10 and 5/12, 6K; 5/15 and 5/VME, 14K; 5/25, 21K; 5/40, 48K; 5/60, 64K; 5/250, 384K; 5/30, 32K.

Although the total I/O capabilities for the various PCs in the series differ, all support analog, ac, dc, high-speed counter, positioning, PID, and ASCII I/Os. Total points are the following: 5/10, 5/12, 5/VME: 512; 5/15: 1,024; 5/25 and 5/30: 1920; 5/40: 2048; 5/60: 3072; and the PLC-5/250: 4096.

The PLC-5 series can be programmed with ladder language and a high-level language via computer. The series has the same local area network capabilities as the PLC-3 series.

Analogic Corporation

This firm offers the DCS 9200 programmable controller, which features 16M of UVPROM. It offers 300 total I/O points (analog, ac, dc, high-speed counter, PID, and ASCII). It is programmed with a Boolean-type language via computer. Its networking capabilities include remote I/O, master-slave, and peer-to-peer formats. The DCS 9200 can perform mathematical functions and basic diagnostics.

Aromat Corporation

Aromat Corporation markets five lines of programmable controllers. The smaller PCs are the M2R, the M1T/M2T, and the FP1. The larger PCs are the FP3 and the FP5.

The M2R and M1T/M2T PCs offer 2500 program steps of EPROM and RAM, with a scan rate of 4.25 microseconds per step. The M2R unit has 16 I/O points (eight input, eight output), but with one expansion I/O board, the total increases to 36 points (20 input, 16 output). The M1T/M2T unit offers 32 I/O points (16 input, 16 output). With four expansion I/O boards, the total I/O for the M1T/M2T increases to 192 (112 inputs, 80 outputs). These I/Os are analog, ac, dc, and high-speed counters.

Both PCs can be programmed with ladder language, either manually or with a computer. Relay ladder logic features of these PCs include 252 internal relays, 64 timers, 48 counters, 32 jump instructions, and 32 master control relay (MCR) instructions. Neither are suitable for use in local area networks, but both perform basic mathematical functions and diagnostics.

The FP1 programmable controller is available in three versions: the C16, C24, and the C40. Memory sizes vary: C16, 900 steps; C24 and C40, 2720 steps. The memory is EPROM and RAM. Base unit I/O points also vary and are given by the number following the C (e.g., the C16 has 16 I/O points). With expansion modules, total I/O can be expanded to 32, 72, and 120 for the C16, C24, and C40, respectively. I/Os are analog, ac, dc, high-speed counter, positioning, PID, and ASCII.

The FP1 can be programmed with either relay-ladder language or Boolean. The software provided with the FP1 allows programming in Boolean, but documentation in relay-ladder language. It can be programmed manually or with a computer. Ladder/Boolean features of the FP1 include data registers, internal relays, timers/counters, and MCRs. The number of each vary among the three versions.

The FP1 is suitable for use in master-slave and peer-to-peer formats, with a personal computer or larger PC (such as the FP3 or FP5) as the master. A C-Net adapter is needed for networking, and communication is through a twisted-pair cable (RS-485). Other communication, such as between the FP1 and a printer, uses the RS-232 interface. The FP1 performs mathematical functions and diagnostics.

The FP3 and FP5 programmable controllers are the largest PCs offered by Aromat Corporation. Both have EPROM and battery-backed RAM, with a scan rate of 0.5 microseconds per step. The FP3 offers 9727 steps of memory, expandable to 15,871 steps, while the FP5 offers 15,871 steps.

The FP3 offers 288, 480, or 768 total I/O points, while the FP5 offers either 1280 or 2048 (analog, ac, dc, high-speed counter, positioning, PID, and ASCII).

Regarding programming language and instruction set, the FP3 and FP5 are like the FP1. As noted previously, both can be used as masters in local area networks with smaller PCs.

ASC Systems

ASC Systems offers four programmable controllers: the ASC A/I, ASC/88, P-C/386, and P-C/486. All use EEPROM and RAM, with memory sizes as follows: ASC/88, 512K; ASC A/I and P-C/386, 4M; P-C/486, 16M.

Total I/O capabilities vary, as follows: ASC/88, 512; ASC A/I and P-C/386, 1024; P-C/486, 2048. These capabilities include analog, ac, dc, high-speed counter, positioning, PID, and ASCII.

All but the P-C/486 can be programmed with relay-ladder language, and all are programmable with Boolean and high-level languages. Programming is performed manually, by CRT, or by computer. All four have networking capabilities, including remote I/O, master-slave, and peer-to-peer formats. All but the P-C/486 adhere to the MAP protocol. All four perform mathematical functions and diagnostics.

Automatic Timing & Controls Company, Inc.

Automatic Timing & Controls produces the ATCOM 64 programmable controller. This PC has 8K of battery-backed CMOS RAM and optional EEPROM (2K and 8K versions are available). Scan times range from 7 to 20 milliseconds. The basic unit has 32 I/O points, expandable to a total of 64. I/Os can be analog, ac, dc, high-speed counter, positioning, PID, and ASCII.

The programming language for the ATCOM 64 is a proprietary language called SNAP™. Programming can be performed manually or with a computer. Communication between the computer and the ATCOM 64 requires an RS-232 interface.

The ATCOM 64 contains 85 timers and 85 counters. It performs basic mathematical functions (addition, subtraction, multiplication, division, comparisons).

Bailey Controls Company

Bailey Controls of Wickliffe, Ohio, offers eight programmable controllers. All of these PCs have battery-backed RAM, and UV-EPROM.

The CLC04 programmable controller has 2K of memory and 13 I/O points (analog, dc, and positioning). It can be programmed in relay-ladder and Boolean languages, manually, by CRT, or by computer. It can be used in a local area network and can perform mathematical functions and diagnostics.

The CSCO has total memory capacity of 144K, and total I/O capability of 14 points. It is otherwise similar to the CLC04, except that it offers remote I/O capabilities.

The CBCO is similar to the CSCO, except it offers 28 I/O points and does not offer positioning I/O capabilities.

The LMM02 programmable controller has total memory capacity of 40K and a total I/O capability of 1024 points (ac, dc, high-speed counter, and PID). It can be programmed in relay-ladder and Boolean languages, manually, by CRT, or by computer. The LMM02 can be used in a local area network and can perform mathematical functions and diagnostics.

The MPC01 programmable controller has a total memory capacity of 148K, and a total I/O capability of 1024 points (ac, dc, high-speed counter, and PID). It is programmed in the same languages, by the same devices, and has the same LAN capabilities as the LMM02.

The MFP01, MFP02, and MFP03 programmable controllers are quite similar. They have total memory capacities of 320K, 512K, and 2512K, respectively. All three PCs have total I/O capabilities of 10,000 points (analog, ac, dc, high-speed counter, PID, and remote). They can be programmed in relay ladder and Boolean languages, manually, by CRT, or by computer. They are capable of use in networks in master-slave or peer-to-peer formats. All three use the MAP protocol and perform mathematical and diagnostic functions.

Basicon, Inc.

This Portland, Oregon, firm produces the SBC-64 programmable controller. It has 8K of RAM and 16K of EPROM and a total of 64 I/O points (analog, dc, ac, high-speed counter, and ASCII). It is programmed with a proprietary language via a computer. The SBC-64 can be used in networks as a slave with a computer as the master.

Blue Earth™ Research

This Mankato, Minnesota, firm produces the Micro-440 programmable controller. This PC offers 32K of battery-backed CMOS RAM, with an additional 128K available through the addition of a memory expansion module. The unit features 14 digital I/O points, but an additional 24 points can be added with an I/O expander module (analog, dc, high-speed counter, and ASCII). The Micro-440 is programmed in BASIC, via CRT or computer. It is suitable for use in master-slave or peer-to-peer formats. The Micro-440 performs basic mathematical functions and diagnostics.

Robert Bosch Corporation

This Charleston, South Carolina, firm offers Bosch CL500 and the Bosch CL300 programmable controllers. Both feature RAM and EEPROM memories. The CL500 has a maximum memory size of 256K words, while the CL300 offers either 16K or 32K of memory. Both offer a total of 4K I/O (analog, ac, dc, high-speed counter, positioning, and ASCII). They are programmable in relay-ladder and Boolean languages via CRT or computer. Both PCs are usable in networks, in either master-slave or peer-to-peer formats. Both adhere to the MAP protocol, and both perform basic mathematical and diagnostic functions.

B & R Industrial Automation

This Roswell, Georgia, firm offers five programmable controllers: MINICONTROL, MIDICONTROL, MULTICONTROL CP40, MULTICONTROL CP60/80, and M264. These PCs have EEPROM and RAM memories. The MINICONTROL, MIDICONTROL, and MULTICONTROL CP40 offer 16K of memory, while the MULTICONTROL CP60/80 and the M264 both offer 42K of memory.

Total I/O points vary among the line: MINICONTROL, 180; MIDICONTROL, 192; MULTICONTROL CP40, 1025; MULTICONTROL CP60/80, 1536; and M264, 264. I/O capabilities include analog, ac (except for the MINICONTROL), dc, high-speed counter, positioning, PID, and ASCII. They are programmed with relay-ladder and Boolean languages, by CRT or computer. They are capable of use in networks in master-slave or peer-to-peer formats. All perform basic mathematical functions and diagnostics.

Cegelec Industrial Controls

Cegelec is an English manufacturer of programmable controllers. It features the MICROGEM 60, GEM 10, GEM 16, GEM 80/131, and GEM 80/160 series of programmable controllers. In addition, Cegelec also produces its own industrial computer.

The MICROGEM 60 is a small PC, featuring a memory capacity of 2000 instructions. The memory is EPROM and battery-backed RAM. The basic unit contains 20 digital I/O points, but with the addition of seven expansion blocks, the total can be increased to 160 I/O points (ac or dc). The total of 160 I/O points is distributed as 96 inputs and 64 outputs. The MICROGEM 60 is programmed in relay-ladder language, either by computer or the MICROGEM programmer. The scan rate is 5 milliseconds per 1000 instructions, and the instruction set includes 32 timers, 32 counters, and 128 internal relays. The MICROGEM 60 is suitable for use in peer-to-peer formats and performs basic diagnostics.

The GEM 10 programmable controller features a memory of 400 instructions with EEPROM and battery-backed CMOS RAM. Its scan rate is 8 microseconds per instruction. The unit offers a total I/O capability of 80 points (analog, ac, and dc). It can be programmed in relay-ladder language or GRAFCET via computer, portable offline programming console, or handheld unit for on-site programming. The instruction set includes counters and timers. Its networking capabilities include both master-slave and peer-to-peer formats. The GEM 10 performs basic mathematical functions and diagnostics.

The GEM 16 programmable controller features a memory size of 800 instructions (EEPROM and battery-backed CMOS RAM). The scan rate is 8 microseconds per instruction. The unit offers a total I/O capability of 256 points (analog, ac, and dc). The programming languages are relay-ladder and GRAFCET. Programming is accomplished via computer, portable offline programming console, or handheld on-site unit. The GEM 16 is suitable for use in master-slave and peer-to-peer format networks, performs basic mathematical functions and diagnostics.

The GEM 80/131 and GEM 80/160 series of programmable controllers are larger PCs. The GEM 80/131 offers a memory size of 20,000 instructions of battery-backed RAM. The scan rate is 1 millisecond per 1000 instructions on average. It has a total of 512 I/O points (analog, ac, dc, positioning, PID, and ASCII). It is programmed in relay-ladder language by CRT or computer. It is usable with remote I/O, including the GEMSMART intelligent I/O modules. It is usable in local area networks, either in master-slave or peer-to-peer formats, and supports the CORONET local area network. It performs basic mathematical functions and diagnostics.

The GEM 80/160 series of programmable controllers is similar to the GEM 80/131, except it offers 512K of RAM and EPROM and a total of 2048 I/O points.

Cincinnati Milacron

Cincinnati Milacron of Lebanon, Ohio, markets two PCs: the APC-500 Relay, and the APC-500MCL. Both offer CMOS RAM memories, but in different sizes. The APC-500 Relay memory size varies from 32K to 6128K, while the APC-500MCL memory size varies from 32K to 64K.

The APC-500 Relay offers a total of 512 I/O points, while the APC-500MCL offers 2048 I/O points (analog, ac, dc, high-speed counter, positioning, PID, and ASCII). The APC-500 Relay can be programmed only in relay-ladder language, either manually or using a CRT, cassette tape, or computer. The APC-500MCL uses a high-level language, and is programmed with the same devices. Both PCs are compatible with remote I/O interfaces, host computers, and peer-to-peer formats. Both perform basic mathematical and diagnostic functions.

Control Technology Corporation

Control Technology offers three programmable controllers: the 2200XM, the 2600, and the 2600-10. All feature NOVRAM memories in varying sizes (2200XM, 10K; 2600 and 2600-10, 16K). Both the 2200XM and 2600 have a total of 160 I/O points, while the 2600-10 offers 320 I/O points (analog, ac, dc, high-speed counter, positioning, PID, and ASCII). All three are programmed with a high-level language via computer. All three are suitable for use in local area networks, either with a host computer or in the peer-to-peer format. All three perform basic mathematical functions and diagnostics.

Creative Control Systems

This company offers the 86KIS programmable controller, which comes with 40K of PROM and 24K of RAM. It sports a total of 512 I/O points (analog, ac, dc, and ASCII). The 86KIS uses a high-level programming language and is programmed via computer. It is suitable for use in local area networks as a slave. It performs basic mathematical and diagnostic functions.

Digitronics Sixnet

Digitronics Sixnet produces the 10MUX programmable controller, featuring 1M of RAM and EPROM. This PC offers 512 I/O points (analog, ac, dc, high-speed counter, positioning, PID, and ASCII). It is programmed with Boolean language using a computer. It can be used in local area networks in remote I/O, master-slave, and peer-to-peer formats. It performs basic mathematical and diagnostic functions.

Divelbiss Corporation

Divelbiss Corporation offers the Baby Bear Bones, the Bear Bones, the PIC Bear Bones, the Boss Bear, and the HD Bear Bones programmable controllers. All use EPROM memory in varying sizes: Baby Bear Bones, 2K or 4K; Bear Bones, 4K; PIC Bear Bones, 4K, 8K, or 16K; Boss Bear, 128K; and HD Bear Bones, 16K.

These PCs offer a variety of I/O points. The Baby Bear Bones sports a total I/O capability of 26 points (analog, ac, dc, and high-speed counter), while the Bear Bones and PIC Bear Bones offer 249 I/O points (analog, ac, dc, and high-speed counter). The Boss Bear has 256 I/O points (analog, ac, dc, high-speed counter, positioning, and PID), while the HD Bear Bones offers 250 I/O (analog, ac, dc, and high-speed counter).

All but the Boss Bear are programmed in relay-ladder language. The Boss Bear is programmed in a high-level language. Programming is accomplished manually or by computer.

Both the Boss Bear and the HD Bear Bones offer remote I/O capabilities. Additionally, the Boss Bear is suitable for use in peer-to-peer formats. It performs basic mathematical functions. All of the Divelbiss programmable controllers perform basic diagnostic functions.

Eagle Signal Controls

Eagle Signal Controls, an Austin, Texas, division of Mark IV Industries, Inc., offers nine programmable controllers: the MICRO 190, the MICRO 190+, the EPTAK 225, the EPTAK 245, the EAGLE 1, the EAGLE 2, the EAGLE 3, the EPTAK 7000, and the CP 8000.

The MICRO 190 and MICRO 190+ offer 32K of battery-backed RAM and UV-EPROM. Both have a total of 128 I/O points (ac, dc, high speed counter). In addition, the MICRO 190+ offers analog and PID capabilities. Both are compatible with remote I/O. Programming is in relay-ladder and Boolean languages, either manually or by computer. Both are suitable for use with a host computer or in peer-to-peer formats. Both perform basic mathematical and diagnostic functions.

The EPTAK 225 and EPTAK 245 have 16K of battery-backed RAM and a total of 128 I/O points (ac and dc). In addition, the EPTAK 245 has analog and PID I/O capabilities. The programming languages are relay-ladder and Boolean. Programming is accomplished either manually or by computer. Both are suitable for use in local area networks (master-slave or peer-to-peer formats). Both perform basic mathematical functions and diagnostics.

The EAGLE 1, EAGLE 2, EAGLE 3, and EPTAK 7000 programmable controllers offer 48K of battery-backed RAM. I/O capabilities are as follows: EAGLE 1 and EAGLE 2, 896; EAGLE 3 and EPTAK 7000, 2048. I/O capabilities include ac, dc, high-speed counter, and ASCII. The EAGLE 2, EAGLE 3, and EPTAK 7000 also offer analog and PID I/O capabilities. Programming is in relay-ladder language via computer. The EAGLE 3 offers remote I/O capabilities. All are suitable for use with host computers or in peer-to-peer formats, and all perform basic mathematical and diagnostic functions.

The CP 8000 offers 128K of battery-backed RAM and UV-EPROM, with a total of 1536 I/O points. In other respects, it is similar to the EAGLE 3 and EPTAK 7000.

Eaton Corporation

The Cutler-Hammer Products Division of Eaton Corporation, Milwaukee, Wisconsin, markets three programmable controllers: the D100 CRA, D100A, and D200. All have RAM and EEPROM memories in various sizes (D100 CRA, 1K; D100A, 1K; D200, 4K). Total I/O points vary. The D100 CRA has a total of 34 I/O points; the D100A, 80 points (analog, ac, dc, high-speed counter, and ASCII). The D200 offers 224 points (analog, ac, dc, high-speed counter, PID, and ASCII). All are programmed in relay-ladder language via computer. In addition, the D200 can be programmed manually. The D100 CRA is suitable for use with a host computer, while the D200 is compatible with remote I/O and can be used in local area networks. Both the D100 CRA and D200 perform basic diagnostic functions. The D200 performs mathematical functions.

Entertron Industries

Entertron Industries of Gasport, New York, offers five PCs: the SK 1600-R SA, SMART-PAK, SK 1800, SK 1600-R, and SK 1600. All feature 8K of battery-backed RAM and EPROM.

Total I/O points vary among the line. The SK 1600-R SA offers 56 total points (analog, ac, dc, high-speed counter, and ASCII). The SMART-PAK offers 22 total I/O points (dc, high-speed counter, and ASCII). The SK 1800 has a total of 88 points (ac, dc, high-speed counter, and positioning). The SK 1600-R has 56 dc I/O points, and the SK 1600 sports 64 ac and dc I/O points.

All are programmed with relay-ladder language using a computer. All perform basic mathematical functions and diagnostics.

Festo Corporation

This Hauppauge, New York, firm markets three programmable controllers: the FPC 101B, FPC 101AF, and FPC 405. All three use EPROM and RAM memories (FPC 101B and FPC 101AF, 12K; FPC 405, 128K). Total I/O capabilities vary: FPC 101B, 35; FPC 101AF, 45; FPC 405, 2272.

All three PCs support dc, high-speed counter and ASCII I/Os. The FPC 101AF and FPC 405 also support analog and positioning I/Os. In addition, the FPC 101AF is compatible with PID I/O modules.

All three use relay-ladder language and Boolean and are programmed via computer. Only the FPC 405 offers remote I/O capabilities. All three are suitable for use with a host computer and perform basic mathematical functions and diagnostics.

Furnas Electric Company

This Batavia, Illinois, company offers four programmable controllers: the PC/96 Plus, 96MFM20, 96MFM28, and the 96/M series.

Both the PC/96 Plus and the 96/M series offer 4K of RAM and EPROM. Total I/O capabilities include 480 points total for the PC/96 Plus (analog, ac, dc, ASCII, remote) and 160 for the 96/M series (analog, ac, dc, high-speed counter, positioning, ASCII, and remote). Both are programmable in relay-ladder language. The instruction set for the 96/M series is expanded. Programming is performed with a computer. Both are useful in local area networks and perform mathematical functions and diagnostics.

The 96MFM20 and 96MFM28 PCs sport 7.6K of RAM and EPROM. I/O points are designated by the last two numbers: 20 for the 96MFM20 and 28 for the 96MFM28 (analog, dc, high-speed counter, positioning, ASCII, and remote). These PCs, like the 96/M series, are programmable in relay-ladder language with an expanded instruction set. Programming is accomplished manually, by cassette tape, or by computer. They have the same networking, mathematical, and diagnostic capabilities as the 96/M series.

GE Fanuc Automation

The GE Fanuc Automation company of Charlottesville, Virginia, is a joint venture of General Electric and Fanuc, Ltd., of Japan. It offers a broad range of programmable controllers that can be classified by series numbers. These are the Series One™, Series Three™, Series Five™, Series Six™, and Series 90™.

The Series One PC is a compact PC for use as relay replacer. It is so small that it easily fits into a shoebox. Its memory size ranges to 1724 16-bit words. Memory availability can be specified as follows: CMOS RAM, 700 words; CMOS RAM, 1724 words; EPROM, 1724 words. The scan rate depends upon the number of words. A scan rate of 12 milliseconds (ms) is available for 250 words, 20 ms for 500 words, 40 ms for 100 words, and 60 ms for 1700 words. The CMOS RAM is battery-backed, using a lithium battery with a two-year lifespan.

The Series One offers a basic 64 I/O points that can be expanded to 112 by adding two additional racks. These are discrete I/O only. The Series One supports remote I/O, with transmission rates of 19.2 Kbaud over 3000 feet. Remote I/O can be added at up to three locations, with 32 I/O points maximum at each location. Communication with remote I/O uses twisted pair wires. Data communication uses the RS-422 serial interface, also at a rate of 19.2 Kbaud.

The Series One PC is programmed with a relay-ladder–based Boolean language. The instruction set includes relays, latch and unlatch, timers, counters, and MCR. Programming is accomplished with a handheld programmer or a Workmaster™ Programmable Control Information Center. The Workmaster™ is a portable, industrial-grade IBM personal computer. Programs can also be loaded from a cassette tape player.

The Series One works well in a local area network as a slave PC. GE Fanuc offers the GEnet Factory Local Area Network, which uses a broadband bus and token passing. GEnet complies with the MAP protocol and uses the RS-422 serial interface.

The Series One Junior PC is a low-cost compact version of the Series One. It is designed for control applications involving 4 to 60 relays. The Series One junior is contained in one package that includes 24 I/O points, 700 words of memory, and a built-in 2-kHz high-speed counter. The 700 words of CMOS memory can be replaced by 700 words of EPROM, if desired. I/O capacity for the Series One junior can be expanded to a total of 64 I/O points by using the standard Series One rack and discrete I/O modules. The Series One Junior PC uses the same programming language, programmers, cassette tape recorder, printer interface, and data communications as the Series One PC.

The Series One Plus programmable controller is an upgraded version of the Series One, yet still small enough to fit into a shoebox. It offers a total of 168 I/O points. Its memory is the same as the Series One. Scan rates, however, are faster (8 ms/500 words, 12 ms/1000 words, and 15 ms/1700 words).

The Series One Plus, unlike the Series One and Series One Junior, is compatible with analog I/O. Twenty four of the total I/O can be used for analog applications. In addition to the basic ladder language, the Series One Plus offers advanced data operations, such as direct interfacing with binary-coded decimal (BCD) devices. Like the Series One, the Series One Plus can be used in the GEnet Local Area Network. The Series One Plus uses the same programming devices as the Series One.

The Series Three PC is still quite compact, but it has greater I/O capability, memory, and functionality. The basic Series Three has 4K of CMOS RAM user memory or 4K of EPROM user memory and 256 I/O points. The system can be expanded to 400 I/O points with the addition of two base units. Of these I/O points, all 400 can be used for discrete and 24 can be used for analog I/O. The Series Three Programmer is built into the basic unit.

Series Three programming is similar to Series One programming. Functions include relay, latch and unlatch, MCR, timer, counter, arithmetical functions, compare functions, and subroutines. The diagnostic functions of the Series Three Programmer help to locate sources of errors in programs.

Scan rates per kilobyte of memory vary. Typically, the scan rate for 1K is 12 ms; for 2K, 20 ms; and for 4K, 35 ms.

The Series Three is suitable for use in local area networks (including GEnet). Remote I/O can be distributed in seven locations over a distance of 3000 feet. The RS-422 serial interface is required.

The Series Five programmable controller is the newest addition to the GE Fanuc Automation line. The Series Five is a compact, modular PC designed for ease of use in programming. It offers a wide range of features compared to other PCs in the same I/O range. The Series Five has 20K of battery-backed CMOS RAM with an additional 24K of EPROM or 4K of EEPROM. It has 2K of I/O points (1K of inputs, 1K of outputs). The programming language is LogicMaster 5, a high-level, expanded relay-ladder logic language. The instruction set includes relays, arithmetic functions, MCR, SKIP, and Boolean operations. Programming is performed via a Workmaster or computer.

The Series Five PC is compatible with a variety of specialized I/O modules. These include digital and analog inputs and outputs, high-speed counter, and ASCII modules. It performs basic mathematical functions and diagnostics.

The Series Six family of programmable controllers includes the Model 60, Model 600, Model 6000, and the Series Six Plus. The capacity of each model (in terms of memory and I/O) increases as the model numbers increase. The Series Six Plus, which has the same amount of memory as the Model 6000, has a greater number of I/O points than the Model 6000. All use ladder language and the same programming devices.

The entire Series Six family uses CMOS memory modules. These modules incorporate lithium batteries to retain memory contents in the event of a power disruption. All memory is measured in 16-bit words. The Model 60 has 4K of memory; the Model 600, 8K; the Model 6000 and the Plus, 64K.

Total I/O capability for the series is as follows: Model 60, 2000; Model 600, 2000; Model 6000, 4000; Plus, 32,000. All of these can be used for discrete I/O, with a smaller number used for analog I/O. The entire Series Six is capable of PID control and uses the same programming devices as the Series Three. The network capability of the Series Six is also the same as the Series Three, except that the Series Six is also compatible in master-slave formats. Communication requires either the RS-232C or RS-422 serial interface.

The Series 90 programmable controllers include the 90-20, 90-30, and 90-70. The 90-20 has 2K of RAM and EEPROM, with a total of 28 I/O points (analog, ac, dc, and high-speed counter). Programming is with LogicMaster 90, a powerful ladder logic/Boolean language. Programming can be performed manually, by CRT, or by computer. The Series 90-20 performs basic mathematical functions and diagnostics.

The Series 90-30 PC has a memory ranging from 6K to 16K of RAM and EEPROM, with 320 to 1000 total I/O points (analog, ac, dc, high-speed counter, positioning, PID, and ASCII). It can be programmed in the same way as the 90-20. In addition, with the use of a 9030 SLCP unit from Adatek, the 90-30 can be programmed in Adatek's State Language. The 90-30 offers remote I/O, master-slave, and peer-to-peer networking capabilities. It, too, performs mathematical functions and diagnostics.

The Series 90-70 ranges in memory size from 32K to 512K of battery-backed RAM. Total I/O points range from 1K to 12K (analog, ac, dc, high-speed counter, positioning, PID, and ASCII). It, too, uses LogicMaster 90, and, with a 9070 SLCP unit, can be programmed in Adatek's State Language. It has remote I/O and networking capabilities. The Series 90-70 adheres to the MAP protocol and performs basic mathematical and diagnostic functions.

Giddings & Lewis Electronics Company

Giddings & Lewis Electronics Company of Fond du Lac, Wisconsin, manufactures five PCs: the PiC 0, PiC 4.9, PiC 49, PiC 409, and PiC 900. All have RAM and EPROM memories.

The PiC 0 offers 2K of memory and a total of 128 I/O points (analog, ac, dc, high-speed counter, and positioning). It is programmed in relay-ladder/Boolean language, manually or by computer. It performs basic mathematical functions.

The PiC 4.9 has 68K of memory and a total of 40 I/O points (analog, ac, dc, high-speed counter, positioning, and ASCII). It is programmed in relay-ladder language by computer. Its networking capabilities include remote I/O, host computer, and peer-to-peer formats.

The PiC 49 and PiC 409 offer 132K of memory, but different I/O capabilities (PiC 49, 256; PiC 409, 2048). I/O types include analog, ac, dc, high-speed counter, positioning, and ASCII. The programming language is relay-ladder, and programming is accomplished via computer. Both are capable of supporting remote I/O and performing in local area networks.

The PiC 900 is the largest in the Giddings & Lewis line, with 512K of memory and 3168 I/O points (analog, ac, dc, high-speed counter, positioning, PID, and ASCII). It is programmed in relay-ladder language by computer. It supports remote I/O, networks, and performs basic mathematical and diagnostic functions.

HMW Enterprises, Inc.

This Pennsylvania firm offers the MPC11 line of programmable controllers. Memory sizes vary between 62K words and 1.3M of memory (RAM and EPROM). The number of I/O points can be either 256 or 512. The programming language is relay-ladder, and programming can be accomplished manually, or by CRT, cassette tape, or computer. All the MPC11 line support remote I/O and have networking capabilities. All perform basic diagnostic functions, and most perform basic mathematical functions.

Honeywell, Inc.

The Industrial Automation and Control Division of Honeywell in Phoenix, Arizona, markets the 620 family and the S9000™ programmable controllers. The 620 family offers CMOS RAM memories in memory sizes that range from 2K to 32K. Total I/O points also vary from 256 to 2048 (analog, ac, dc, high-speed counter, positioning, PID, and ASCII). Programming is in relay-ladder language via CRT or computer. Most of the 620 family support remote I/O, and all are suitable for use in local area networks. They all perform basic mathematical and diagnostic functions.

The S9000 family of programmable controllers offers up to 32K of memory and up to a total of 2048 I/O points, digital and analog. Inputs are limited to 96. A wide variety of specialized I/O modules are available, including ac, dc, high-speed counter, positioning, PID, and ASCII.

Programming is in relay-ladder/Boolean language, with an expanded instruction set. It supports remote I/O and has networking capabilities. It, too, performs mathematical and diagnostic functions.

Horner Electric, Inc.

This Indianapolis, Indiana, firm produces the HE 2000 programmable controller. The memory includes 1K of PROM, 32K of battery-backed RAM, and 64K of EPROM. There are 600 total I/O points (analog, ac, dc, PID, and ASCII). The programming language is a high-level ladder language. Programming is accomplished by computer. The HE 2000 supports remote I/O and has networking capabilities, either with a host computer or in a peer-to-peer format. It performs mathematical and diagnostic functions.

ICON Corporation

This Woburn, Massachusetts, firm produces two motion control programmable controllers: the MC^2 and the C60. The MC^2 features 1024K of RAM and a total of 144 I/O points (analog, ac, dc, high-speed counter, positioning, and ASCII). Programming is in relay-ladder language, either manually, or by CRT or computer. It is capable of performing basic mathematical and diagnostic functions.

The C60 features 2K of EEPROM and 16K of RAM, with a total of 24 I/O positioning points. It is programmed in Boolean language, manually or by cassette tape.

Idec Corporation

Idec Corporation of Sunnyvale, California, offers three programmable controllers: the Micro 1, the FA-2 Junior, and the FA-3S. The Micro 1 has 0.6K of EEPROM; the FA-2 Junior, 4K of RAM, EPROM, and EEPROM; and the FA-3S, 1K or 4K of RAM, EPROM, and EEPROM. The Micro 1 has 28 I/O points, while the other two have 256. The Micro 1 offers ac and dc I/O capabilities, while the FA-2 Junior and the FA-3S offer analog, ac, dc, high-speed counter, positioning, and remote I/O capabilities. In addition, the FA-3S offers PID and ASCII I/Os. Programming languages are relay ladder and Boolean. Programming is performed manually, by cassette tape, or by computer. All three are suitable for use with a host computer or in peer-to-peer formats. All perform basic diagnostic functions, and the FA-2 Junior and FA-3S perform basic mathematical functions.

Industrial Indexing Systems

Like ICON Corporation, Industrial Indexing Systems offers programmable controllers for motion control applications. The product line includes the VME-1000 single-axis servo-motion controller, the MM-5 single-axis servo-motion control system, the MM-10-T and MM-T-Plus single-axis servo-motion controllers, the ServoPro™ single-axis motion control system, the MSC-250 and MSC-850/32 programmable multi-axis servo-motion controllers, and the VSC-850 multi-axis servo-motion controller. These programmable controllers are dedicated to motion control. Additional information on the features of these programmable controllers can be obtained from the manufacturer, whose address is given in appendix A.

International Parallel Machines

This New Bedford, Massachusetts, firm manufactures the IP1610 programmable controller and the IP1612 family of programmable controllers. All offer 500 steps of EEPROM memory and a total of 28 I/O points. Most of these machines support analog, ac, dc, high-speed counter, and ASCII I/Os. The programming language is relay-ladder, and programming is accomplished manually or by computer. Some of these machines are suitable for use with host computers. They all perform basic mathematical and diagnostic functions.

Klockner-Moeller

This Franklin, Massachusetts, firm markets the PS 3, PS 306, PS 316, and PS32 German-made programmable controllers. All have RAM and EEPROM memories. The PS 3 is a small, low-cost programmable controller. It offers 3.6K of memory and 152 I/O points (analog, ac, dc, and high-speed counter). It is programmed in a relay-ladder/block diagram language, manually or by CRT, cassette tape, or computer. The language set includes logic operations (AND, OR, XOR), basic arithmetical functions, jumps, timers, counters, and comparator operations. It is suitable for use with remote I/O modules, a host computer, or in a peer-to-peer format. It performs basic diagnostic functions.

The PS 306 is a compact, high-speed programmable controller, offering 32K words of RAM and 8K words of EEPROM. It supports a total of 333 I/O points (analog, dc, high-speed counter). The programming language is relay-ladder/block diagram, and programming is accomplished by computer. The programming instruction set is similar to that of the PS 3, as is its networking capabilities.

The PS 316 is a high-speed industrial programmable controller that features 32K words of RAM and 32K words of EPROM. Total I/O points are 4624 (analog, ac, dc, high-speed counter, positioning, and ASCII). It is otherwise similar to the PS 306 in its capabilities.

The PS32 PC line features up to 64K of RAM and 64K of EPROM. It offers a total of 4992 I/O points (analog, ac, dc, high-speed counter, positioning, and ASCII). The programming language is relay-ladder/block diagram, and it is programmed via computer. It offers 64 OFF timers, 64 ON timers, 128 counters, 128 adders, 128 multipliers, and 128 dividers. It is suitable for use with remote I/O and in networks.

Klockner-Moeller offers the SUCONET Field Bus local area network. This proprietary local area network is suitable for use with all Klockner-Moeller PCs and features polling between master and slave and half-duplex transmission.

Minarik Electric Company

Minarik Electric Company offers the MicroMaster LS programmable controller. This device has 3K of RAM and PROM and a total of 120 I/O points (ac, dc, high-speed counter). The programming language is Minarik SPC Control Language, an enhanced form of ladder logic. Programming is accomplished manually or by cassette tape or computer. The MicroMaster LS is suitable for use with a host computer and performs basic mathematical and diagnostic functions.

Mitsubishi Electronics America, Inc.

This Mt. Prospect, Illinois, firm produces the F and A families of programmable controllers. The F family includes the F1/F2, the FX0, and the FX PCs. The FX0 is the smaller PC, with 0.8K of EEPROM and 30 dc and high-speed counter I/O points. It is programmed with relay-ladder and Boolean languages manually or by CRT or computer. It performs mathematical and diagnostic functions.

The F1/F2 features 2K of RAM and EPROM with a total of 120 I/O points (analog, ac, dc, high-speed counter, positioning, and ASCII). It uses relay-ladder and Boolean languages and is programmed manually, by CRT, or by computer. It is suitable for use with a host computer or in peer-to-peer formats and performs mathematical and diagnostic functions.

The FX PC features 8K of EEPROM memory and 256 I/O points. It has the same type of I/O capabilities, uses the same programming languages and devices, and features the same mathematical and diagnostic functions as the F1/F2.

The A family of PCs features a range of RAM and EPROM memory sizes from 8K to 448K and a variety of I/O points (256 to 2048). These points are analog, ac, dc, high-speed counter, positioning, PID, ASCII, and remote I/Os. The family uses the same programming languages, is programmed by the same devices, and features the same networking capabilities as the FX PC. The A family adheres to the MAP protocol.

Modicon, Inc.

Modicon, Inc., of North Andover, Massachusetts, produces a full-size PC family, designated the PC-984. The 984 comes in four versions: the compact version (984-120 to 984-145), the slot-mounted version (984-380 to 984-785), the chassis-mounted version (984-A to 984-X), and the Mod Connect version (AT-984, MC-984, Q-984, and V-984).

The compact version offers between 1.5K and 8K of CMOS RAM with a total of 1024 I/O points. Programming is in relay-ladder language via CRT or computer. These PCs are suitable for use with a host computer. They all perform basic mathematical and diagnostic functions.

The slot-mounted version offers between 6K and 32K of CMOS RAM and EPROM, with I/O points ranging from 1024 to 65,536. Programming is in relay-ladder language or by CRT or computer. Some of these PCs are compatible with remote I/O, and all are suitable for use in local area networks with a host computer. Some adhere to the MAP protocol.

The chassis-mounted version offers between 8K and 32K of CMOS RAM and between 32,768 and 65,536 I/O points. Programming is in relay-ladder language by CRT or computer. All are compatible with remote I/O and usable in local area networks. They all adhere to the MAP protocol.

The Mod Connect PCs offer between 12K and 16K of CMOS RAM and a total of 7168 I/O points. Programming is in relay-ladder/block diagram language via CRT or computer. All the PCs in this series are compatible with remote I/O and are suitable for use in local area networks.

Omron Electronics, Inc.

This Schaumburg, Illinois, firm produces the SP-10, C20K and C20H series, C200H, C500, C1000H, and C2000H programmable controllers. All feature RAM and EPROM memories. The SP-10 has 144 words of memory, with 10 ac and dc I/O points. It is programmed in relay-ladder/Boolean language. It can be used in peer-to-peer formats and performs basic mathematical and diagnostic functions.

The C20K and C20H series of PCs feature between 1194 and 2.6K words of memory with 140 to 160 I/O points. They are programmed in relay-ladder/Boolean language manually or by CRT, cassette tape, or computer. They are compatible with remote I/O and can be used with a host computer. Both perform basic mathematical and diagnostic functions.

The C200H, C500, C1000H, and C2000H offer a variety of memory sizes, from 7K words to 36K. I/O points vary from 512 to 2048 (analog, ac, dc, high-speed counter, positioning, PID, and ASCII). The programming language is relay-ladder/Boolean, and programming is performed manually or by CRT, cassette tape, or computer. The networking capabilities include remote I/O, master-slave, and peer-to-peer formats. The C1000H and C2000H adhere to the MAP protocol.

Opto 22

This Temecula, California, firm markets five programmable controllers: the Mistic 100, Mistic 150, Mistic 200, LC4, and LC2. The Mistic 100 contains 64K of RAM and 32K of ROM with 4096 I/O points. As with the other Opto 22 programmable controllers, the I/Os include analog, ac, dc, high-speed counter, positioning, PID, and ASCII. The Mistic 100 is programmed with a high-level language via CRT or computer. As with the other Opto 22 PCs, it is compatible with remote I/O and is suitable for use in local area networks in either master-slave or peer-to-peer formats. It can perform mathematical and diagnostic functions.

The Mistic 150 contains 256K of RAM and 2M of ROM, with 7500 I/O points. It is programmable in a high-level language by computer. It offers mathematical and diagnostic functions.

The Mistic 200 contains 4M of RAM and 4M of ROM and a total of 18K I/O points. The LC4 has 64K of RAM and 32K of ROM with a total of 4096 I/O points. The LC2 offers 32K of RAM and 32K of ROM with a total of 1024 I/O points. These are programmed with a high-level language via computer.

The Opto 22 programmable controllers are capable of using Adatek's State Language.

Pep Modular Computers, Inc.

This Pittsburgh, Pennsylvania, firm produces four programmable controllers: the IUC 9000, VME 9000, VME 9030, and VME 9040. The IUC 9000 offers 1M of ROM, 1M of RAM, and 64K of EEPROM with a total of 924 I/O points. It is programmable in relay-ladder language via computer and is suitable for use in master-slave and peer-to-peer format networks. It performs mathematical and diagnostic functions.

The VME 9000 offers 1M of ROM, 1M of RAM, and 64K of EEPROM, with a total of 416 I/O points (analog, dc, high-speed counter, positioning, and ASCII). It is programmable in relay-ladder language by computer. It is suitable for use in master-slave and peer-to-peer formats and performs mathematical and diagnostic functions.

The VME 9030 has 2M of ROM, 5M of RAM, and 256K of EEPROM. The VME 9040 has 2M of ROM, 4M of RAM, and 256K of EEPROM. In other respects, these two are similar to the VME 9000.

Phoenix Contact, Inc.

This Harrisburg, Pennsylvania, company offers the Interbus series of PCs, Interbus-R, Interbus-S, and Interbus-C. These PCs have a variety of RAM (96K to 512K) and I/O points that vary in size from 29 to 4096 (analog, ac, dc, high-speed counter, positioning, and ASCII). Programming is in relay-ladder language with an augmented instruction set via computer. These PCs are compatible with remote I/O and local area networks. They perform basic diagnostic functions.

Precision Microcontrol Corporation

This San Diego, California, company offers the DCX and DC2 lines of programmable controllers for motion control applications. They feature 128K of RAM and ROM and 96 to 144 I/O points. They are programmed with a high-level language via CRT or computer and offer remote I/O and host computer networking capabilities.

Pro-Log Corporation

This Monterey, California, company markets the Busbox/IPLC, which features 96K of RAM and 600 I/O points. It is programmed by CRT or computer in relay-ladder language, and offers networking capabilities.

Reliance Electric Company

This Cleveland, Ohio, firm offers three lines of programmable controllers: the DCS line, the AutoMate line, and the Shark line.

The DCS line of PCs includes the DCS 5000 and the AutoMax DCS. These PCs have between 80K and 512K of battery-backed RAM, with 12K of I/O points (analog, ac, dc, high-speed counter, positioning, PID, ASCII, and remote). They are programmed in ladder language by computer and are suitable for use in local area networks.

The AutoMate line features between 1K and 104K of EEPROM, RAM, and NOVRAM, with 64 to 8192 I/O points. They are programmed in ladder language, support remote I/O, and perform mathematical and diagnostic functions.

The Shark line contains the Shark X and Shark XL, with 1K or 2K of EEPROM and 60 to 160 I/O points (analog, ac, dc, and high-speed counter). They are programmed in relay-ladder/Boolean language manually or by cassette tape or computer. The XL is suitable for use in local area networks.

Siemens Industrial Automation, Inc.

The Programmable Controllers Division of this Alpharetta, Georgia, company offers the SIMATIC line of PCs. This includes the 100U, 115U, 135U, and 155U, and the TI line taken over from Texas Instruments.

All the Siemens PCs have RAM and EPROM memories. The original Siemens line (100U, 115U, 135U, and 155U) displays memory capacities from 20K to 896K and I/O points from 448 to greater than 1M (analog, ac, dc, high-speed counter, positioning, PID, ASCII, and remote). Programming languages are relay-ladder/Boolean and GRAFCET. These PCs are suitable for use in local area networks, and all but the 100U adhere to the MAP protocol.

There are quite a few PCs in the TI series (TI315, TI330, TI330S, TI425, TI430, TI435, T1525, TI535, TI545, TI560T, TI565T, and the T1575). These PCs feature memories ranging in size from 700 to 512K. Total I/O points also vary from 96 to 8192. Programming is in ladder language and GRAFCET, and programming is accomplished by computer. These PCs are suitable for use in local area networks and perform basic mathematical and diagnostic functions.

SKH Systems, Inc.

This Medina, New York, company produces two programmable controllers: the SKH-19 and SKH-23. Both have EPROM and RAM memories, ranging in size from 32K to 64K. The SKH-23 sports 28 I/O points, while the SKH-19 offers 256 I/O points (analog, ac, dc, high-speed counter, and ASCII). The SKH-19 also offers positioning and PID I/O. These PCs are programmed in relay-ladder language by computer and perform mathematical and diagnostic functions.

Square D Company

The Automation Products Division of the Square D Company (Milwaukee, Wisconsin) offers the SY/MAX series of programmable controllers.

The MICRO-1 is a small programmable controller featuring 600 steps of EEPROM and a total of 28 ac and dc I/O points (16 inputs, 12 outputs). The MICRO-1 is programmed in relay ladder/Boolean language, and programming is accomplished either manually or by computer. It performs basic diagnostic functions.

The SY/MAX Model 50 is another small PC with 1K to 4K steps of battery-backed CMOS RAM. An additional 4K steps of EPROM or EEPROM are available. It provides a total of 256 I/O points (analog, ac, dc, high-speed counter). The programming language is relay-ladder/Boolean, and programming is accomplished either manually or by computer. The SY/MAX Model 50 can be used with a host computer and performs mathematical and diagnostic functions.

The Model 300 features up to 2K of RAM or UV-PROM memories with a scan rate per kilobyte of 30 ms. A mixture of RAM and UV-PROM is also available as an option. This U.S.-made PC has 256 I/O points. It is programmed with ladder language either manually or by CRT, cassette tape, or computer. It uses the RS-422 serial interface. The Model 300 has high-speed counting, positioning, PID control, and remote I/O capabilities. It can be used with the SY/NET local area network. It offers mathematical and diagnostic capabilities.

The SY/MAX Model 400 PC features 4K, 8K, or 16K of battery-backed RAM or UV-PROM. It offers a total of 4K I/O points (analog, ac, dc, high-speed counter, positioning, PID, ASCII, and remote). The programming language is relay ladder, and programming is accomplished manually or by CRT, cassette tape, or computer. It can be used with the SY/NET local area network and performs mathematical and diagnostic functions.

The Model 500 has up to 8K of RAM or RAM/UV-PROM memories with a scan rate of 2.6 ms per kilobyte. Total I/O capabilities include up to 2000 points. It features all the functions of the Model 300 (control, networking, etc.) with additional advanced capabilities. These advanced capabilities include expanded mathematical functions (square root, random number, absolute value, etc.), scan control (GOTO, subroutines, etc.), matrix operations, immediate communications update, Boolean functions (AND, OR, XOR), etc. It has PID control and positioning capabilities and uses the SY/NET local area network. The Model 500 is programmed the same as the Model 300 (manually, CRT, cassette tape, or computer).

The SY/MAX Models 600 and 650 PCs feature 16K or 26K of RAM and a total of 8K I/O points (analog, ac, dc, high-speed counter, positioning, PID, ASCII, and remote). They are programmed in relay ladder language, either manually, by computer, cassette tape, or CRT. They can be used with the SY/NET local area network.

The Model 700 features 64K of RAM memory. It offers a memory control unit (MCU) that includes nonvolatile backup memory (bubble memory). Scan rate per kilobyte of memory is 1.7 ms. The Model 700 has 8K total I/O points. The ladder language instruction set includes relays, latch and unlatch, MCR, timers, counters, data comparison, binary to BCD, BCD to binary, mathematical operations (including square root, absolute value, sine and cosine, base 10 logarithms, natural logarithms, power functions, and advanced trigonometric operations), PID control, and Boolean operations. It is also capable of motion control (positioning). It can be programmed manually or by CRT, cassette tape, or computer. Like the other SY/MAX models, the Model 700 uses the SY/NET Local Area Network.

Systems Engineering Associates, Inc.

This Arvada, Colorado, firm markets the S3000 and M4000 programmable controllers. The S3000 has 52K of RAM and EPROM, and a total of 256 I/O points (analog, ac, dc, high-speed counter, positioning, and ASCII). It is programmed in relay ladder language by computer. It performs mathematical and diagnostic functions.

The M4000 has 26K of CMOS RAM and a total of 48 dc and high-speed counter I/O points. In other respects, it is like the S3000.

Telemecanique, Inc.

Telemecanique, Inc., of Westminster, Maryland, markets a variety of programmable controllers. These include the TSX-17, TSX-47, TSX-67, TSX-87, and TSX-107.

The TSX-17 features 24K of RAM and EPROM memories with 160 I/O points. It uses ladder, Boolean, and GRAFCET programming languages. It can be programmed manually or by CRT or computer and offers mathematical and diagnostic functions.

The TSX-47 offers 32K to 112K of RAM and EPROM memories. It has up to 1024 I/O points. It is capable of high-speed counting and positioning operations. It uses ladder, Boolean, and GRAFCET programming languages. It can be programmed manually or by cassette tape or computer. The TSX-47 is suitable for use in peer-to-peer local area networks and performs mathematical and diagnostic functions.

The TSX-67-40 has 224K RAM and EPROM memories. It has 2048 I/O points and is programmable with ladder, Boolean, and GRAFCET languages. It offers high-speed counting and PID control capabilities. It is suitable for use in local area networks and with remote I/O. It too performs mathematical and diagnostic functions.

The TSX-87-40 and the TSX-107-40 have 352K of RAM and EPROM memories. They offer 2048 I/O points and are otherwise similar to the TSX-67.

Telemonitor

TeleMonitor, of Herndon, Virginia, offers the Master*Link PC. This device has 64K of RAM and ROM and I/O capabilities that include analog, ac, dc, high-speed counter, positioning, PID, ASCII, and remote. Programming is with a high-level language manually or by CRT or computer. It performs mathematical and diagnostic functions.

Toshiba International Corporation

This Houston, Texas, firm manufactures the EX and M series of programmable controllers. The EX series includes the EX14B, EX100, and EX500. These PCs have battery-backed RAM and EEPROM memories ranging in size from 1K to 8K and ranging in total I/O points from 12 to 512. I/O capabilities include analog, ac, dc, high-speed counter, and positioning. In addition, the EX500 offers PID and ASCII I/O capabilities. Programming is in relay-ladder language by computer. All but the EX14B offer remote I/O capabilities, and all these PCs are suitable for use with a host computer.

The M series includes the M20 and M40. These PCs offer 3K or 4K of RAM and EEPROM and 20 or 40 I/O points. The I/O capabilities are the same as the EX100. Programming is in relay-ladder language via computer. Both feature remote I/O, are suitable for use with a host computer, and perform basic mathematical and diagnostic functions.

Triconix Corporation

Triconix of Irvine, California, offers the TRICON programmable controller. It has 350K to 1M of RAM memory and features 2528 I/O points. It is programmable with ladder and high-level languages. It is capable of PID control and is programmed by CRT or computer. The TRICON is suitable for use with remote I/O and in local area networks. It can perform mathematical and diagnostic functions.

UTICOR Technology, Inc.

UTICOR Technology of Bettendorf, Iowa, produces three PCs: the Director 4001, Director 4002, and Director 6001. The Director 4001 has 6K of CMOS RAM and is capable of controlling up to 256 discrete I/O points. It is compatible with a range of specialized I/O modules: analog, high-speed counter, stepper motor driver, thermocouple input, remote I/O, and data modules. It also provides eight PID control loops. The programming language is relay ladder, and programming is accomplished with a program loader. It performs basic mathematical, Boolean, and diagnostic functions. Communication with remote I/O employs the RS-422 twisted pair cable, while communication with printers employs the RS-232 interface.

The Director 4002 is similar in functionality to the 4001. The base unit comes with 64 discrete I/O points, expandable to 256 with expansion I/O modules.

The Director 6001 features larger memory capacity and total I/O capabilities (256K of CMOS RAM and 4096 I/O circuits). It is functionally similar to the 4001.

Westinghouse Electric Corporation

The Distribution and Control Business Unit of Westinghouse (Pittsburgh, Pennsylvania) offers three series of programmable controllers: the 50 Series, 500 Series, and 2000 Series. The 50 Series includes the PC-50 and PC-55 programmable controllers. The PC-50 is a small, inexpensive PC offering 2K words of battery-backed RAM with a scan rate of 2 msec/K. The basic unit offers 10 digital inputs and six relay outputs. With the addition of six expansion I/O modules, total I/O capacity can be expanded to 64 I/O points. It is compatible with the 500 Series I/O modules, providing tremendous flexibility (digital, analog, counter, timer, comparator, and ASCII communications). All the Westinghouse PCs are programmed in relay-ladder language or in a text-based language called Statement List. All the Westinghouse PCs are compatible with the L1 local area network, a Westinghouse data highway. The PC-50 performs basic mathematical and diagnostic functions.

The PC-55 is a more advanced, yet small, PC featuring 8K words of battery-backed RAM. On-board I/O includes 16 digital inputs, 16 digital outputs, eight analog inputs, one analog output, four interrupt inputs, and two high-speed counter inputs. When combined with the 500 Series I/O modules, total I/O capability is expanded to 288 points.

The 500 Series is a mid-sized series of PCs, featuring up to 10K words of battery-backed RAM (or EPROM or EEPROM). Maximum I/O points number 256 (up to 32 500-Series I/O modules).

The Westinghouse 2000 Series PCs offer up to 48K words of RAM, and a total of 2048 digital and 128 analog I/O points. This series also employs the 500-Series I/O modules. It offers networking with the entire Westinghouse line of PCs using the L1 local area network.

Wizdom Systems, Inc.

This Naperville, Illinois, firm markets six PCs: the SPLIC JR, SPLIC, 186 Coprocessor, 386 Coprocessor, Controller I, and Controller II.

The smaller PCs are the SPLIC JR and SPLIC, which feature 128K of RAM and I/O totals of 64 and 80, respectively. As with all Wizdom Systems PCs, I/O capabilities include analog, ac, dc, high-speed counter, positioning, PID, and remote. All Wizdom PCs are programmed in relay-ladder language by computer and offer networking capabilities. The two coprocessors and two controllers offer 128K of RAM, and a total of 4096 I/O points.

Yaskawa Electric America, Inc.

This North Brook, Illinois, company offers the GL40, GL60, and GL70 series of PCs. The GL40 has 8K words of program memory, 2K words of data memory, and 2K words of memory dedicated to motion control. It features 2056 I/O points. As with the other two product lines, these include analog, ac, dc, high-speed counter, positioning, PID, and ASCII. Programming language is relay-ladder, and programming is accomplished manually or by CRT or computer. The GL40 has networking capabilities and performs basic mathematical and diagnostic functions.

The GL60 and GL70 are larger PCs. The GL60 features 32K words of program memory and up to 32K words of data memory. The GL70 has 64K words of program memory and 32K words of data memory. All these PCs have 12,288 I/O points and are programmed in relay-ladder language or GRAFCET. They are suitable for use in local area networks, either with a host computer or in peer-to-peer formats, and perform mathematical and diagnostic functions.

Z-World Engineering

This Davis, California, company produces a series of PCs programmable in Dynamic C™, a version of the C computer programming language. These include the Little Giant™, the Tiny Giant, the Little PLC™, and the SmartBlock™.

The Little Giant features 512K of battery-backed RAM and 256K of EPROM with a total of 39 I/O points (analog, dc, high-speed counter, and ASCII). It is suitable for use in networks. The Tiny Giant is similar to the Little Giant, except that it offers 26 I/O points and no high-speed counter I/Os.

The Little PLC is a miniature controller offering up to 512K bytes of battery-backed RAM and up to 512K of EPROM. (It comes from the factory with 32K of each.) It offers 16 I/O points (eight inputs and eight outputs). It employs the RS-485 half-duplex asynchronous port.

The SmartBlock is a microprocessor core module, and is thus not strictly a PC. It contains up to 256K of EPROM and up to 512K of battery-backed RAM. An additional 512 to 2048 bytes of EEPROM are available.

Summary

The preceding data on programmable controllers should provide you with a broad overview of the diversity of PCs available today. The data presented in this chapter should be confirmed with the manufacturer, as changes in the programmable controller industry occur quite rapidly.

Domestic Suppliers of Programmable Controllers

ABB Process Automation, Inc.
Division of Asea Brown Boveri, Inc.
650 Ackerman Rd.
Columbus, OH 43202
(614) 261-2000

Active Systems Group, Inc.
P.O. Box 3249
Redmond, WA 98073-3249
(206) 885-0200

Adatek, Inc.
700 Airport Way
Sandpoint, ID 83864
(208) 263-1471

Allen-Bradley Co.
1201 S. 2nd St.
Milwaukee, WI 53204
(414) 382-2000

Analogic Corp.
Measurement Controls Division
8 Centennial Dr.
Peabody, MA 01960
(508) 977-3000

Aromat Corp.
629 Central Ave.
New Providence, NJ 07974
(908) 464-3550

ASC Systems
P.O. Box 2523
North Canton, OH 44720
(216) 499-1210

Automatic Timing & Controls Co., Inc.
King of Prussia, PA 19406
(215) 337-5500
(800) 441-8245

Bailey Controls Co.
29801 Euclid Ave.
Wickliffe, OH 44092
(216) 585-8500

Basicon, Inc.
14273 NW Science Park Dr.
Portland, OR 97229
(503) 626-1012

Blue Earth Research
310 Belle Ave.
Mankato, MN 56001
(507) 387-4001

Robert Bosch Corp.
P.O. Box 10347
Charleston, SC 29411
(803) 552-6000

B&R Industrial Automation Corp.
1325 Northmeadow Parkway, S-130
Roswell, GA 30076
(404) 772-0400

Cegelec Industrial Controls
Kidsgrove, Stoke-on-Trent
Staffordshire ST7 1TW
England
In the U.S., call Mr. Peter Wyatt at
(215) 651-0707

Cincinnati Milacron Co.
Electronic Systems Division
Mason Rd. and State Route 48
Lebanon, OH 45036
(513) 494-5319

Control Technology Corp.
25 South St.
Hopkinton, MA 01748
(508) 435-9595

Creative Control Systems, Inc.
4413 Fernlee
Royal Oak, MI 48073
(313) 549-1121

Digitronics Sixnet
P.O. Box 767
Clifton Park, NY 12065
(518) 877-5173

Divelbiss Corp.
9776 Mt. Gilead Rd.
Fredericktown, OH 43019
(614) 694-9015

Eagle Signal Controls
A Division of Mark IV Industries, Inc.
8004 Cameron Rd.
Austin, TX 78753
(512) 837-8300

Eaton Corp.
Cutler Hammer Products
Programmable Controllers
4201 North 27th St.
Milwaukee, WI 53216
(800) 833-3927 Ext. 1000

Entertron Industries, Inc.
3857 Orangeport Rd.
Gasport, NY 14067
(716) 772-7216

Festo Corp.
395 Moreland Rd.
Hauppauge, NY 11788
(516) 435-0800

Furnas Electric Co.
1000 McKee St.
Batavia, IL 60510
(708) 879-6000

GE Fanuc Automation
P.O. Box 8106
Charlottesville, VA 22906
(804) 978-5000

Giddings & Lewis Electronics Co.
666 S. Military Rd.
Fond du Lac, WI 54935
(414) 921-7100

HMW Enterprises, Inc.
604 Salem Rd.
Etters, PA 17319
(717) 938-4691
(717) 765-4690

Honeywell, Inc.
Industrial Automation and Control
16404 N. Black Canyon Highway
Phoenix, AZ 85023
(602) 863-5000

Horner Electric, Inc.
1521 E. Washington St.
Indianapolis, IN 46201
(317) 639-4261

ICON Corp.
26 Connecticut St.
Woburn, MA 01801-5662
(617) 933-9666

Idec Corp.
Programmable Controllers Group
1213 Elko Dr.
Sunnyvale, CA 94089-2211
(408) 747-0550

Industrial Indexing Systems
626 Fishers Run
Victor, NY 14564
(716) 924-9181

International Parallel Machines, Inc.
700 Pleasant St.
New Bedford, MA 02740
(617) 990-2977

Klockner-Moeller
25 Forge Parkway
Franklin, MA 02038
(508) 520-7080

Minarik Electric Co.
901 East Thompson Ave.
Glendale, CA 91201-2011
(818) 502-1528

Mitsubishi Electronics America, Inc.
800 Biermann Ct.
Mt. Prospect, IL 60056
(708) 298-9223

Modicon, Inc.
An AEG Company
One High St.
North Andover, MA 01845-2699
(508) 794-0800

Omron Electronics, Inc.
Factory Automation Systems Division
1 East Commerce Dr., Dept. T
Schaumburg, IL 60173
(708) 843-7900
(800) 826-6766

Opto 22
43044 Business Park Dr.
Temecula, CA 92590
(714) 695-9299
(800) 321-6786

PEP Modular Computers, Inc.
750 Holiday Dr. No. 9
Pittsburgh, PA 15220-2783
(412) 279-6661

Phoenix Contact, Inc.
P.O. Box 4100
Harrisburg, PA 17111
(717) 944-1300

Precision MicroControl Corp.
8122 Engineer Rd.
San Diego, CA 92111
(619) 565-1500

Pro-Log Corp.
2555 Garden Rd.
Monterey, CA 93940
(408) 372-4593

Reliance Electric Co.
P.O. Box 17438
Cleveland, OH 44117
(216) 266-7000

Siemens Industrial Automation, Inc.
Programmable Controllers Division
Dept. SD
100 Technology Dr.
Alpharetta, GA 30202
(404) 740-3000
(800) 964-4114

SKH Systems, Inc.
2611 County Line Rd.
Medina, NY 14103-9411
(716) 735-7350

Square D Co.
Automation Products Division
P.O. Box 472
Milwaukee, WI 53201-9732
(414) 332-2000

Systems Engineering Associates, Inc.
P.O. Box 750
14989 West 69th Ave.
Arvada, CO 80001
(303) 421-0233

Telemecanique, Inc.
2002 Bethel Rd.
Westminster, MD 21157
(410) 581-2000

TeleMonitor
512 Herndon Parkway
Suite G
Herndon, VA 22070
(703) 437-5201

Toshiba International Corp.
Industrial Division
13131 W. Little York Rd.
Houston, TX 77041
(713) 466-0277

Triconex Corp.
15091 Bake Parkway, Suite D
Irvine, CA 92718
(714) 768-3709

UTICOR Technology, Inc.
4140 Utica Ridge Rd.
Bettendorf, IA 52722-1327
(319) 359-7501

Westinghouse Electric Corp.
Distribution and Control Business Unit
5 Parkway Center
Pittsburgh, PA 15220
(412) 937-6547
(800) 525-2000

Wizdom Systems, Inc.
1260 Iroquois Ave.
Naperville, IL 60563
(708) 357-3000

Yaskawa Electric America, Inc.
3160 MacArthur Blvd.
North Brook, IL 60062-1917
(708) 291-2340
(800) 633-5756

Z-World Engineering
1724 Picasso Ave.
Davis, CA 95616
(916) 757-3737

B

State Logic Control
A paper on the theory behind State Logic
Reprinted with permission of Adatek

This paper will introduce you to the world of State Logic control.

State Logic is a different approach to the control and management of machines, systems, and processes offering real and powerful advantages over traditional control techniques.

State Logic is a powerful and proven control methodology. State Logic is built from different fundamental principles than Relay Ladder Logic, Boolean combinational logic, and Basic or other computer programming languages sometimes used to control machines and processes.

Adatek State Logic control was designed specifically for today's more complex and challenging control requirements. Using State Logic for control tests only the user's knowledge of the system to be controlled, not computers, programming, codes, or theories.

Overview of the Popular Control Strategies Currently in Use

To bring perspective to State Logic, we must briefly examine current control approaches.

Controllers

Today's industrial controllers were spawned by different and once totally separate branches of industrial automation. Programmable (Logic) Controllers, or PLCs, were designed to control discrete parts manufacturing machines and systems. Process controllers were designed to manage continuous process operations. While both resulting products are computers with nearly matching electronic components, they differ significantly in implementation because of their different roots. Their differences lie almost entirely in their physical packaging and their software.

Today the worlds of process control and discrete parts manufacturing control are merging. Most modern controllers offer some combination of discrete, process, motion, batch, or data handling capabilities.

Control equipment suppliers from each camp are now adding modifications to their products in an effort to accommodate the much broader needs of today's industrial control. In most cases, however, the controllers or computers simply cannot be tweaked sufficiently to serve well the requirements of the other control applications. This usually results in a control system that fits one area of control adequately but is cumbersome in use and thin on capability when solving control problems other than those from its originally chosen field.

The Programmable (Logic) Controller. The PLC was invented in 1969 as a replacement for hardwired panels of electro-mechanical relays, switches and related devices. The PLC offers the speed, I/O scanning, and factory floor toughness required for reliable performance in industrial real-time applications.

Because the PLC was originally intended only to replace relay type functions, the PLC's software architecture was designed with an embedded model of the relay panel. In Relay Ladder Logic (RLL) an arrangement of contacts and coils in an electrician's ladder diagram schematic format is used to implement a combinational (or Boolean) logic structure.

In combinatorial logic for control, the status of an output is determined by the status of a certain combination of inputs. Combinatorial logic offers little facility for describing the status of an output in relation to time, to the operational flow of the process under control or to the other outputs.

Special function boxes are used in RLL for control requirements that cannot be implemented with an arrangement of contacts. Clever contact arrangements are used to create the interlocking logic required for events that occur sequentially in the process.

A combinatorial logic foundation with an RLL representation was an excellent choice at the time PLCs were introduced because the job of the PLC at that time was to replace only the functionality of hardwired relay control panels with a computer substitute. Hardwired relays are limited to combinational logic because of practical limitations inherent in a system physically bounded by wires and switches. RLL remains an excellent option today for small single-sequence relay replacement applications.

As the application range of the PLC expanded, the limitations of RLL in a digital environment began to show. As users became more computer literate, they realized that the PLC is actually a powerful computer and therefore not limited to the same physical or practical limitations of hardwired systems. Many modifications have been implemented to permit the PLC to do things that relay panels could not do but that computers can. Ultimately it is the programmer who is called upon to compensate for these limitations, either by creating complex, convoluted, and often unmanageable control programs or by limiting actual system design to only that which can be implemented practically and within schedule using RLL.

RLL provides a framework for control that forces the programmer to focus on each output individually rather than the flow and operation of the system under control. RLL programming requires the use of many special case solutions for accomplishing common control needs, eliminating the possibility of clean, direct or elegant programming. The latching techniques required to cause concurrently operating activity sequences to interact in a ladder diagram model result in a program that is very difficult to modify or troubleshoot.

The process controller. Process control really consists of two distinct parts. One part is the modulating control that must be performed on a continuous basis. This is process loop control employing feedback to maintain process variables at the desired setpoint.

The roots of this type of control strategy are in the pneumatic controllers that were first used at the end of World War I. In the 60's, the pneumatic controllers were replaced by solid state controllers that implemented electronically the same modulating control that had been implemented pneumatically. In the late 70's, the microprocessor-based controllers began replacing the solid state controllers.

These controllers used digital techniques to model and emulate the modulating control that had been previously achieved by hardwiring together individual amplifiers and other components. This was accomplished in software. The basic hardware was similar to a PLC, except because of the need to handle analog variables, these controllers stressed a different type of interface and floating point math.

In addition to modulating control, these controllers also had to provide discrete and sequencing logic capability. Every process passes through different states as it operates, requiring different functions to occur in the modulating control, such as operating manually or automatically and starting up or shutting down the process.

Originally relays and switches had been used to implement that logic with the pneumatic controllers. The solid state controllers had replaced relays with AND/OR gates that were wired together. The microprocessor based controllers simply implemented those AND/OR gates in software using a block language.

As PLC users also began using process controllers, they acquainted the process control vendors with the PLC ladder logic concepts. Eventually process controllers offered both relay ladder logic and the AND/OR gate language.

Thus, while the process controller is ideally suited for half its requirements, modulating control, it inherently carries the same drawbacks and shortcomings of the PLC.

The computer

Computer is a category name that can be applied to a wide range of devices used on the factory floor, ranging from a business or personal type IBM PC pressed into industrial service to a DEC MicroVAX. For control applications, these computers contain a CPU to execute instructions, an interface to instruments and actuators through process I/O, and an interface to a user through a CRT and keyboard. These are also the components of the PLC or process controller.

The difference between the industrial computer and the PLC or process controller really is in their software and packaging. The PLC and process controller have software only for a specific job. The industrial computer comes with only a general shell software, the operating system such as DOS or VMS, and tools to build other software, such as Assembler or C compilers.

Because of its general nature, the industrial computer can be made to perform more functions than the PLC or process controller. Graphic displays, recipe handling, and data logging are typical examples.

In the area of complex plant start-up or machine sequencing, programs written in C can implement more complex logic than a PLC. However, the user must be knowledgeable in computer programming to accomplish this logic. The industrial computer is a general tool and its languages are computer-oriented, not process-oriented.

Many software packages are on the market that convert a computer into either a PLC or a process controller. While they often add enhancements such as graphics or data storing, these packages still contain the fundamental limitations in implementing logic of PLCs or process controllers.

Up until now, implementing sequencing and complex logic with an industrial computer required a computer software expert.

The State Engine Alternative

The control approaches just covered can usually be made to function in most control applications but often require very clever programming or compromises in the true potential of the system under control.

A *state engine* offers an alternative choice to traditional controllers or computers. A state-engine-based controller or computer allows the implementation of the powerful State Logic control strategy. This approach requires little of the compromise often required when choosing between PLCs, Process Controllers, and Computers. State engines also offer new capabilities and a level of programming efficiency not previously available with any other control methodology.

A state engine is the complete firmware code set that, when embedded in a control hardware platform, supports the execution of a State Logic type of control methodology.

In Adatek's implementation of State Logic control, a "State Engine" resides in the control hardware. The State Engine is available in several controller and computer hardware platforms. The State Engine software has an embedded State Logic model or "template" into which a description of any control application will naturally fit.

Adatek also provides extremely easy-to-use program development tools, such as ECLiPS, for describing the desired control system. ECLiPS run on any IBM PC or compatible where the program is developed and then downloaded to the State Engine in the control hardware. Adatek control products require no previous programming experience or the attendance of schools or classes.

The Theory behind Adatek's State Logic Control

Adatek developed state logic for the control and management of physical systems with any mix of discrete, process, batch, data, motion control or other. To accomplish this, Adatek developed a control approach rooted in the field of Finite State Machine Science. The principles embodied in Finite State Machine Science are ideal for modeling any problem that will ever be encountered in control.

Finite State Machine Science has become the strategy of choice in many other disciplines, such as electronics design, but industrial control has trailed behind other industries in tapping the power of this approach. The beauty of finite state machine thinking is that it simplifies implementation while increasing power and flexibility. Adatek's State Logic control offers a means of harnessing the power of Finite State Machine Science in all automation applications.

State Logic as a very high-level programming language

State Logic is a methodology for control not based on combinational logic but on Finite State Machine Theory. State Logic is so powerful because it inherently contains a template or framework perfect for modeling any real world process. The fact is, all physical processes are described using State Logic principles, first in the system design process and then again during the design of the control system.

In large part, the unique power and flexibility of State Logic flows from faithfulness of the match between the problem to be solved and the model upon which it is based. The State Logic model fits a control application that involves discrete and analog inputs, operator interface, data logging, and highly sequential operation, as well as RLL fits a job replacing a few relays or a word-processing program fits the job of writing a letter.

All Very High Level Languages (VHLL) have embedded in them a model of the problem they are designed to address. Lotus 123 embeds a model of an accountant's columnar pad and uses terms like "row" or "column." Relay Ladder Logic uses an electrician's ladder schematic model and uses terms like "rungs" and "contacts." State Logic uses a finite state machine model and uses terms like "Tasks" and "States."

VHLLs make the job of program development, program modification, and system management dramatically easier and faster than using lower level languages. Their purpose is to shift the burden on the user away from understanding computers and codes to simply understanding the problem they wish to solve. Another positive impact of a VHLL is that it shifts the user's focus to "top-down" or systems-level thinking and problem solving.

One problem with VHLLs is that they only work well with a single class of problem. The opposite end of the language spectrum, assembly code, can solve virtually any problem type but takes enormous programming time and requires years of computer and programming study.

The problem to be solved and the model embedded in the VHLL must be a good fit. Even small discrepancies between the problem to be solved and the embedded model in the VHLL can cause inefficiency in program development and ambiguity in the program. You could probably write a letter with a spread-sheet program or make a chart with a word-processing program. You can, given time and patience, handle data or control highly sequential applications with Relay Ladder Logic. However, in each of these cases the mismatch between the embedded model and the real application will cause inefficiency and ambiguity.

A key to understanding how State Logic works is to first understand the State Logic model or template embedded in every State Engine. Once this template is well visualized, it is not difficult to imagine how any control problem of any complexity can be "loaded" into the template using the program development tools provided in any Adatek programming tool.

Further, a good understanding of the underlying structure of State Logic will also allow you to see why advanced process diagnostics can be easily developed and why program modifications are so straightforward.

The State Logic model

All real-world processes move through sequences of states as they operate. Every machine or process is a collection of real physical devices. The activity of any device can be precisely described as a sequence of steps in relation to time. For example, a hydraulic cylinder can exist in only three states: extending, retracting, and at rest. Any desired behavior for that cylinder can be described by a sequence of these three states. The off/on cycles of a digital point can be precisely described as a series of states changing over time. Even a continuous process goes through start-up, manual, run, and shutdown phases.

All physical activity can be precisely described in this way. It is also not difficult to imagine that the test condition or event that causes the cylinder or digital output to change states can be explicitly expressed (for example, "if the temperature is over 100° F, turn on the warning light and go to shutdown procedure").

Time and sequence are natural dimensions of the State model just as they are natural dimensions of the design and operation of every control system, process, machine and system. State Logic Control uses these attributes as the components of program development. As a result the control program is a complete and clear snapshot of the system under control.

The State Logic model is a hierarchy that consists of *tasks*, subdivided by *states*, which are described by *statements*.

TASKS: The primary structural element of the State Logic model. A Task is a description of process activity expressed sequentially and in relation to time. If we were describing an automobile engine, the Tasks would include the Starting System Task, the Fuel System Task, the Charging System Task, the Electrical System Task, and so on.

All but the most simple machines and processes will contain multiple Tasks operating in parallel. This concurrency is necessary since the process or machine to be controlled must do more than one thing at a time. For maximum power, flexibility and programming economy a control system must be able to match this simultaneous operation of otherwise independent activity streams within every process. State

Logic provides for the modeling or programming of many Tasks that are mutually exclusive in activity yet interactive and joined in time.

Typical control Task types include: motion task, alarm task, drilling task, diagnostic task, filling task, start-up task, heating task, mixing task, conveyor task, failure forecast task, data logging task, shutdown task, operator interface task, inspection task, report building task, etc.

It is valuable to remember that State Logic does not invent concurrently executing Tasks. Tasks are an inherent part of the operation of the machine or process. State Logic simply uses this structure as the model for program development. The essence of State Logic is that it takes full advantage of this natural pattern of operation inherent in every physical system.

Segmenting an overall process into individual Tasks is not a rigid process. The user is offered great flexibility in creating Task architectures that fit a specific design philosophy, application requirement, or—perhaps most important—to reflect the personal style of the program developer.

States: The building blocks of a Task. The activity of a Task is described as a series of steps called States. Each State is a chunk of time through which the Task activity passes. The amount of time it takes a State to complete its activity is not fixed in the State Logic model but is wholly dependent upon the specification of the actions ascribed to that State by the user.

A State describes the status or value of an output or group of outputs. These are the outputs of the control system and are therefore the inputs to the process. A single State may describe the status of many outputs and the status of those outputs may be linked to one or many input conditions and variable values.

Every State will also contain the rules that allow the Task to transition to another State. A State may offer a collection of input conditions and variables that can cause a change to another State. A State can be directed to move to any other State anywhere in the Task sequence. Input conditions and variable values from anywhere in the process can be used to control what State will be the destination at transition.

A State then, is a subset of a Task that describes the output status and the conditions under which the Task or process will change to another State.

The States of a Task taken in aggregate provide an explicit description of that sequence of activity of the process or machine under control. Further, it provides a unambiguous specification of how that aspect of the process will respond in all conditions. Adatek offers State Engines with either 32 or 256 concurrently executing Tasks and each Task can use up to 99 States to describe its activity.

Statements: The user's command set for creating State descriptions. The desired output related activity of each State can be described by using *Statements*. Statements can simply initiate a direct action or can base an output status change on a conditional statement or a combination of conditions. Any input value or variable can be used in conditional Statements. Variables can include State status from other Tasks as well as typical integer, time, string, analog, and digital status variables. An unlimited number of Statements can be used in a State.

Adatek's State Logic provides the tools for describing any machine or process action. Rather than invent an arbitrary and proprietary code scheme, State Logic uses the same normal English terms, phrases or sentences already used to describe these events. The user may use any terminology to describe their process including their choice of terms to describe control actions, I/O names or variable names. Adatek programming software will then sort everything out automatically and generate the precise, finished code understood by the controller to execute the user's control strategy.

An Adatek State Engine has an embedded model into which specific control applications can be described. The model is organized by 32 or 256 (depending on model) concurrently executing Tasks. The activity for each Task can be subdivided into up to 99 States. Each State can be described with an unlimited number of Statements.

Thus, State Logic provides a methodology for developing a control program by describing the desired result in a logical, intuitive manner. This approach to control encourages the program developer to think about program development from the operational flow of the system under control and tightly linked to the principles of design in the system or machine to be controlled.

The result of combining the State Logic model and natural English Statements is startlingly simple. If you can write a description of the operation of your desired system, you have written the control program.

A Few Special Powers of State Logic Control

The fundamental underlying principles upon which State Logic is based always make State Logic control much easier and significantly faster to use. State Logic control programs will always be far easier to modify than other types of control strategies. State Logic programs will always execute (scan) about 10 times faster than RLL programs, given the same application and class of control hardware.

In addition to this, State Logic's inherent structure makes certain capabilities possible that are impossible or impractical in other control approaches. This paper will expand on three of those capabilities.

Advanced on-line process diagnostics

A State Engine based controller can be programmed to have diagnostic capability well beyond anything that can be done in a non-State Logic control approach.

A State Engine embedded in the controller running a State Logic control program will always contain a complete, explicit, and unambiguous snapshot of perfect system operation.

Because of this, it is not difficult to imagine that diagnostics can be written into a State program such that the controller automatically flags aberrations from desired operation. In the event of such an occurrence, a message can be sent or corrective action initiated.

More significantly, such an occurrence can be made to automatically trigger another TASK into action. This Task can be written to diagnose the cause of the problem and report the diagnosis in English. The State Logic structure is ideal for creating real-time advanced process diagnostics as part of the control program. This is done simply by describing what the system should do in the event of an aberration.

State Logic programs inherently contain valuable information about the status or outputs in relation to time and the status of all other outputs in the system.

The combination of these attributes result in the creation of far more sophisticated process diagnostics, created quickly and without any computer programming knowledge.

Data handling

Interfacing with the user and handling data is extremely simple in a State Logic control program. The program describes the exact sequence of operation that is desired. Therefore, it is easy to insert precisely when a message should be sent to the Operator or Maintenance person.

The State Engine processes the instruction to send the message to one of multiple serial communication ports. These can be connected to operator panels, dumb CRTs, data-collecting computers, and smart actuators such as servos or robots, or more. The message is contained within the instruction and can be ASCII text to be displayed or a coordinate for a robot or even something different. A programming tool such as ECLiPS makes setting up the message as simple as typing it.

Reading in data works the same way. Since the State Logic controller program describes the sequence of operation, it is easy to insert instructions to read information when that information is needed and should be available.

The State Engine contains an instruction that will read one of a series of serial communication ports and when information is available place the information into a memory location to be used by other State Logic programs.

Because the State Engine takes advantage of the full power of the computer, this information can be in the form of digital values, real numbers, or ASCII characters. The State Engine controller instructions allow the User to test and manipulate the data in any form equally. It does not impose artificial limits that do not exist in the hardware as do relay ladder logic controllers.

By combining these output and input commands, sophisticated operator interfaces using just dumb terminals can be built including menu selection and data entry. Data can be stored in files and custom reports and logs can be output to printers or terminals.

The State Engine combined with its programming tool, ECLiPS, allows a non-computer oriented user to develop sophisticated interfaces and data processing applications.

Higher levels of tools for the user

Any control programming approach that does not accurately model the machine or process under control will limit the scope and power of the higher level tools that can be created to assist in program development, system commissioning and system management.

State Logic lends itself to extension into higher level support tools due to its simplicity and tight link between its structure and the true sequential nature of physical systems. ECLiPS, Adatek's general purpose programming development and system management tool is a good example.

ECLiPS (English Control Language and Programming System). ECLiPS (English Control Language and Programming System) and other ECLiPS based application specific programs provides the user programming and system management tools for any control hardware that incorporates a State Engine.

ECLiPS is a complete program development, debugging, documentation, and on-line management software package designed specifically for the control of industrial systems, machines, or processes. ECLiPS is a high-level programming environment requiring the programmer or maintainer only to understand the machine or process to be controlled.

Control programs are developed in ECLiPS automatically from a normal English Language description of the desired control system.

ECLiPS has a translator built in that transparently converts an English language description of your control system into precise code that can be downloaded to any State Engine based computer or controller.

ECLiPS includes a runtime module called OnTOP. OnTOP provides maintenance technicians or operators with access to real-time monitoring information and troubleshooting tools.

ECLiPS provides a variety of documentation options. Documentation is created automatically by ECLiPS from information available from the programming method. The user simply fills out a menu to design the right documentation package.

Conclusion

State Logic control is rapidly growing in popularity. Over 2000 Adatek State Engine based controllers are now in service by users such as AT&T, Boeing, DuPont, Ford, IBM, Monsanto, Westinghouse, and the Department of Defense. State Logic's acceptance is also evident from the number of control equipment manufacturers that have added Statelike features or commands to their traditional control approaches. While these features are helpful, they do not permit the controller or computer to approach anywhere near the power, flexibility, or ease of use of a true State Machine.

State Logic empowers the user. Its objective is to become transparent, allowing clear focus on control system design rather than control system programming. State Logic will significantly shorten the distance between the logical and clever thinking of the engineer and the final control program. State Logic will permit the factory floor smarts of the maintenance technician to be applied directly and effectively to getting processes and machines commissioned quickly and increasing uptime over time.

Current control techniques may have been the first approaches for the control machines, systems, and processes with computers, but State Logic will be the last.

Complete Sample Program

Cans are filled with two chemicals and mixed as the cans move down a conveyor belt. When a can is placed on the conveyor—tripping the Can In Place limit switch—and if neither the Fill nor the Mix tasks are currently active, then the conveyor will start. The conveyor will run until a can arrives at either the Fill or Mix stations, at which time the conveyor will stop and wait for another can and any Filling or Mixing tasks to be completed.

When a can arrives at the Fill Station, tripping the Can at Fill limit switch, Chemical Valve 1 will open until the Fill Weight is above 20 lbs. Then Chemical Valve 2 will open until the Fill Weight is above 30 lbs. Then the Fill Station will wait for another can to arrive to begin the cycle again.

When a can arrives at the Mix Station, tripping the Can at Mix limit switch, the Down Mixer motor will run until the Mixer Down limit switch is tripped. The Mixer Motor will then start. After the Mixer Motor has run for 30 seconds, the Mixer Up Motor will run until the Mixer Up switch is tripped. A counter is incremented and the operator notified after each can is mixed.

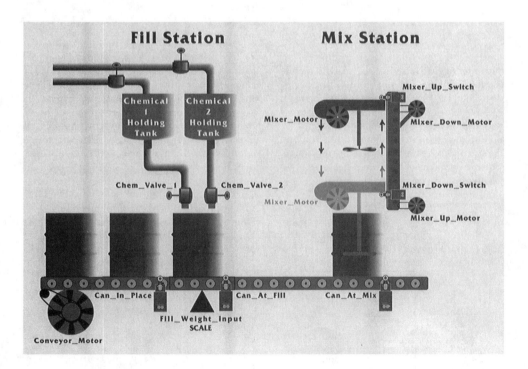

```
PROJECT: BATCHING SYSTEM


Task: Fill_Station

    State: PowerUp
            When Can_At_Fill is on, go to the Batch_Chem_1 State

    State: Batch_Chem_1
            Open Chem_Valve_1
            When Fill_Weight is above 20 pounds, go to Batch_Chem_2.

    State: Batch_Chem_2
            Open Chem_Valve_2 until Fill_Weight is more than 30 lbs.
            then go to the Batch_Complete State.

    State: Batch_Complete
            When Can_At_Fill is off, go to the PowerUp State.

Task: Mix_Station

    State: PowerUp
            If Can_At_Mix is on, go to the Lower_Mixer State.

    State: Lower_Mixer
            Run the Mixer_Down_Motor until the Mixer_Down_Switch is tripped,
            then go to the Mix_Chemicals State.

    State: Mix_Chemicals
            Start the Mixer_Motor.
            When 30 seconds have passed, go to the Raise_Mixer State.

    State: Raise_Mixer
            Run the Mixer_Up_Motor until the Mixer_Up_Switch is tripped,
            then go to the Batch_Complete.

    State: Batch_Complete
            When Can_At_Mix is off, go to the Update_Inventory State.

    State: Update_Inventory
            Add 1 to Can_Inventory
            Write "Batch Cycle Complete" to the Operator_Panel and go
            to PowerUp.

Task: Conveyor

    State: PowerUp
            When Can_In_Place is on and
            (Fill_Station Task is in the PowerUp or Batch_Complete) and
            (Mix_Station Task is in the PowerUp or Batch_Complete),
            go to Start_Cycle.

    State: Start_Cycle
            Start the Conveyor_Motor.
            When Can_In_Place is off, go to Index_Conveyor State.

    State: Start_Cycle
            Start the Conveyor_Motor.
            When Can_In_Place is off, go to Index_Conveyor State.

    State: Index_Conveyor
            Run Conveyor_Motor.
            When Can_At_Fill or Can_At_Mix is on, go to the PowerUp State.
```

STEPS™: Ladder Logic Programming for The StepLadder™ PLC (v2.0)

Reprinted with permission of Active Systems Group, Inc.

Introduction

The StepLadder PLC is a simplified ladder logic device that can be used as a local controller or as a building block in a total control system. STEPS software is designed to make PLC programming and Ladder Logic very natural processes. Systems for controlling machinery and processes can be very complex. STEPS and The StepLadder PLC have been designed to make the pieces of those systems as simple as possible.

Installing The StepLadder PLC

The StepLadder PLC mounts directly upon the 50-pin header type connector that comes on many IO module racks. This connector is designed to accept intelligent modules, like The StepLadder PLC, or ribbon cable connectors. Only the vertical headers (pins perpendicular to the mounting board) can be used. Some IO module racks use a connector with a right angle or 90° bend to the connector pins. This type is *not* compatible with The StepLadder PLC.

Locate pin 1 on the header connector. It is usually marked on the connector itself by a small arrow or triangle that has been molded into the plastic connector housing, or it may be marked on the mounting board with another arrow or triangle, or it might be marked with the number "1." Pin 1 on The StepLadder PLC is clearly marked on top of the circuit board at one end of the 50-pin connector. Once these two pins have been aligned, simply press The StepLadder PLC firmly into the mounting board connector.

Installing STEPS on Your Personal Computer

STEPS requires an IBM Personal Computer or compatible using DOS 2.1 or higher, one floppy disk drive, a serial port, and 512K of memory.

To use STEPS on a computer with a hard disk drive, first create a directory for STEPS. Then simply copy the STEPS disk onto that directory. As a backup precaution, put your STEPS diskette in a safe place.

If you intend to use STEPS from a floppy drive system, make a copy of the STEPS diskette for your use and then store the original in a safe place for backup.

Demo Help on Learning and Using STEPS

Learning to use STEPS only takes a few minutes. If you have installed it on your computer, you might want to start it up and try a few things as you read these instructions. Just change your logged drive and directory to that which you installed STEPS and type STEPS and press Enter. (Type STEPSDMO if you have the demonstration version.) Your STEPS diskette came with some sample programs on it. Feel free to load any of them and make a few changes.

What Is a Program File?

STEPS program files are text files—one or more lines of information that can be printed or displayed. Certain characters in a STEPS Program File have special meaning when they appear in the first (leftmost) position on a line. The most important of these is the less-than character ("<") which marks the line as a Logic Line. All characters that have special meaning when appearing in the first position are listed here:

<   A Logic Line
>   Definition of a timer or counter
@   Assign a name to an IO point
%   Set a communications data rate

All other lines in a program file are comments. This allows extensive documentation of the program because you can put several comment lines above and below each Logic Line.

The Logic Line

A Logic Line defines something that you want the StepLadder to do. This can be called a Step in your program, or a rung in the ladder logic. Your program can have several Logic Lines, depending on the task or tasks that it must perform. All Logic Lines begin with a less-than character ("<"). This is followed by a Logic Definition and an Output Definition.

The Logic Definition defines the logical part of the Step, and the Output Definition defines the place to put the result of the logic. Both of these may contain Point References, which refer to the physical inputs and outputs or to the internal IO points (timers, counters and internal relays) of the StepLadder. The greater-than character (">") marks the beginning of the Output Definition.

Example   `<1.2,3>0`

where

<	Marks the line as a Logic Line
1.2,3	Is the Logic Definition
>0	Is the Output Definition

AND & OR

As you can see from this example, there are other characters that have special significance when contained in the Logic Definition—specifically, the period or dot character, which means "AND," and the comma, which means OR. So, in the previous example, you would say that when IO points 1 and 2 or 3 are ON, then point 0 will be turned ON by the StepLadder. STEPS would draw this for you as:

```
      1           2
 +--! !-----! !--+
 !          3    !                                      0
 +---------! !--+----------------------------------( )
```

The Normally Closed Point

All of the points used in the previous example were Normally Open. That is to say that their normal, or logically false, state is an open contact, and their logically true state is a closed contact. There will be times when you want the StepLadder to take action when the point is NOT ON. This is called a Normally Closed point. You can specify any point in the Logic Definition as Normally Closed by preceding it with the backslash character ("\").

Example   `<1.2,\3>0`

STEPS would draw this for you as shown. Notice that there is a backslash character inside the contact symbol for IO point 3. Thus, when IO points 1 and 2 are on, or IO point 3 is NOT ON, the StepLadder will turn on point 0.

```
      1           2
 +--! !-----! !--+
 !          3    !                                      0
 +---------!\!--+----------------------------------( )
```

Using Parentheses

You can use parentheses to structure your logic step in the same way you would structure a mathematical formula. The logic within the left and right parentheses gets solved and its result is included in the solution for the complete logic line.

We can change the meaning of our example by enclosing the IO points 2 and 3 in parentheses.

Example   `<1.(2,3)>0`

```
                2
              +--! !--+
      1       !  3    !
    +--! !--+--! !--+
                !                                           0
                +------------------------------( )
```

Output Definition

The Output Definition is the part of the logic line that tells the StepLadder what to do with the result of the logic. It begins with the greater-than sign (">") and is followed by any physical IO point or internal IO point designation. Internal points include Internal Relays, Timer run, Timer reset, Counter count and Counter reset.

Any physical IO point or internal relay can be further designated with an Output Mode. This is done by following the IO point name or number with L, R or O. An 'N for 'Normal is assigned when no other Output Mode is designated.

L-Latch   With latch mode the IO point is turned ON and will stay on, even when the result of the logic is no longer true.
R-Reset Latch   Turns off an IO point that has been latched.
O-One Shot   The point is turned on for one complete scan of the logic program.
N-Normal   The point is turned ON when the logic is true and OFF when the logic is false. The default mode.

The StepLadder has 24 physical IO points numbered 0 through 23. There are also 88 Internal Relays, which are numbered 96 through 183. The two Timer Run points are designated as TM0 and TM1; and their respective Timer Resets are TR0 and TR1. Counter Count points are CT0 through CT9 and Counter Resets are CR0 through CR9. See *Timer/Counter operation* for a complete description of timers and counters. The following example shows how to designate output 23 as a one-shot-type output.

Example   `<1.2,3>23 o`

```
      1          2
    +--! !-----! !--+
    !          3    !                                23
    +---------! !--+------------------------------(O)
```

Names for IO Points

You can assign a name to any physical IO point or Internal Relay that you want to use in your program; this makes program printouts much more readable and meaningful.

To do so, you must enter a Point Name Line in your program. A Point Name line begins with the AT character ("@") followed by the name you want to use, then an equals sign, and finally the IO point number.

Example   @DoorA=2
@DoorB=3
@SwDoor=1
@SwOn=4
@DomeLite=0
<SwSoor.(\DoorA,\DoorB),SwOn>DomeLite

This short program controls the dome light in a two-passenger automobile. There are inputs from the dome light switch to indicate the ON position or the DOOR position. Doors A & B have contacts that close when the respective door is closed. STEPS will draw this as follows.

```
                     DoorA
               +--!\!--+
   SwDoor  !   DoorB!
  +--! !--+--!\!--+
  !               !
  !               !
  !               !
  !          SwOn !                              DomeLite
  +----------! !--+------------------------------( )
```

Setting the Communications Data Rate

The StepLadder PLC will communicate via its serial port at 9600 baud (bits per second), unless it has been programmed for an alternate rate. The data rates available are 1200, 2400, 4800, and the default 9600 baud.

To program the StepLadder communications rate enter a line in the program that has a percent sign ("%") in the first character position. Then simply add the desired data rate to the line.

Example   %1200

As soon as the The StepLadder PLC is reset, it will use the newly programmed rate, which might require that you to change the rate at which STEPS communicates.

Timer/Counter Set-Up

Timers and Counters can be used in many different ways. STEPS requires a set-up line in the program to specify each timer and counter in use. The set-up line must begin with the greater-than sign (">") followed by the timer or counter point identification TM0, TM1, or CT0 through CT9, and then three or four parameters (the fourth is optional) separated by spaces. This sets up the timer or counter operation. The four parameters are as follows:

Preset The value set into the time or count when the timer or counter is reset. Maximum 65535 for counters and 655.35 for timers.

Output The IO point number (or name) to be operated when the time or count hits zero.

OutMode The mode of output operation. Valid OutModes are N-Normal, O-One shot, T-Toggle and L-Latch.

Auto Optional parameter to use auto reset when the timer or counter hits zero.

The next example will toggle IO point 2, every 10.5 seconds for as long as IO point 1 is on.

Example   >tm0 10.50 2 t a
              <1>tm0

```
                                            tm0
           1                            +------+   2
        +--! !--------------------------! 10.50!--(T)
                                        !     A!
                                        !      !
                                        +------+
```

Timer/Counter Operation

Timers and counters have much in common. They both count down from their pre-set to zero and they both have the same output modes of Normal, Latch, Toggle and One Shot. A Timer (or Counter) IO point refers to the result of a Logic Line which becomes the Input to the Timer or Counter. Timer IO Points are TM0 and TM1; Counter IO Points are CT0 through CT9.

Running/Counting

Timers run (or count down) as long as the Timer IO Point is ON. Counters will count down one count whenever the Counter IO Point makes a transition from OFF to ON.

Resets/Presets

A Timer Reset (TR0 or TR1) will set the timer to its preset value every 1/100th of a second, as long as the Timer Reset is ON. A Counter Reset (CR0 etc.) will set the counter to its preset value when the Counter Reset makes the transition from OFF to ON. Timer and Counter Resets turn OFF the timer or counter output when it is specified as Normal or Toggle output mode.

Outputs

Timer and Counter Outputs are either physical IO points or internal relays. When used in Normal Output Mode, the IO Point is set ON when the count or time hits zero. The Toggle Mode complements the IO Point so that is goes ON if it was OFF; and OFF if it was previously ON. When Latch Mode is used, the timer or counter sets the IO Point and its Latch to ON. When Latch Mode is used for a timer or counter, the latch can only be reset by a Logic Line used specifically for resetting that IO Point Latch.

Auto

When Auto Reset is used, the timer or counter is set to its preset value any time it hits zero. This feature is very useful when you need to use a timer that runs continuously as a timer base for other counters.

Entering & Editing

STEPS editing is very much like a word processor. Almost all the keys on the keyboard work like you would expect from any text editing program The major difference is the fact that STEPS uses most of the screen to draw logic diagrams for you.

When you select 'E from the main menu, STEPS will put up the editing screen. The top line still shows the current file and now includes the current line number. The cursor is positioned where the current line is displayed. If the line is a Logic Line, then STEPS draws the logic for you. Any time STEPS comes upon a Logic Line, it draws. The best way to get acquainted with STEPS editing is to load one of the program files that come on the STEPS disk and browse through it. Make a few changes and even save the revised file if you wish.

Moving through the file

Down Arrow	Moves to the next line.
Up Arrow	Moves to the previous line.
PgDn	Moves to the next Logic Line.
PgUp	Moves to the previous Logic Line.
Ctrl-End	Moves to the last line in the file.
Ctrl-Home	Moves to the first line in the file.
Ctrl-PgUp	Moves half way to the top.
Ctrl-PgDn	Moves half way to the bottom.

Editing within a line

Right Arrow	Move cursor right one position.
Left Arrow	Moves cursor left one position.
Home	Moves to first position in line.
End	Move to end of line.
Del	Delete character at cursor.
Backspace	Delete character to left of cursor.
Tab	Space to next Tab stop (6,11,16,26,etc.)
Enter	Redraw if line is a Logic Line if pressed only once.

Inserting & deleting lines

Ins	Inserts a line before the current line.
Ctrl-Y	Deletes the current line.

Enter (twice)   Insert a line after the current line when pressed twice.

All printable characters are inserted into the current line. Any other nonprintable characters are ignored, with the exception of 'Escape which returns to the STEPS main menu.

Accessing Program Files

When STEPS begins there is no current, or active, file being edited. You can begin entering your program at this point; or you can load a file from one of the available disk drives on your computer. Once you have loaded a program from disk or saved your program to disk, that program becomes the current file. In order to access any disk files, first you must go to the Files Menu, which is selected by pressing F at the STEPS Main menu.

New File

Use the New File selection on the Files Menu if you wish to clear out the program you are working on and start fresh. This blanks out the program in memory and the current filename. It does not erase anything from your disk. This will put you into the editing mode at line 1 of your new program.

Drive Select

You can use this selection to change the current drive that STEPS uses to look for files. The directory for any drive you select will be the current directory as was selected by DOS when STEPS was begun. In other words, STEPS can select a drive but can only display the current directory for the selected drive (it has no way of changing a drive's current directory). So it is a good practice to make sure that the directory you want to access, on each drive you want to access, is the current directory prior to starting STEPS. However, when opening or saving a file, you can always access any directory by entering the complete DOS path and filename.

When you press D from the Files Menu, the current directory for the current drive is displayed. You may enter a new drive letter if you want to select a different drive on your computer.

To Open a File

Select O from the Files Menu to open a file for editing or for subsequent transfer to The StepLadder PLC. STEPS will display the current directory for the current drive and invite you to enter a filename. You can enter a name from the directory displayed, or a complete path and filename from any directory.

Save the Current File

Select S from the Files Menu to save the current file. This selection is provided to make it easy to save your work at regular intervals and when finished.

Save As

Select A from the Files menu to save the current file and assign a new filename, or to save the file in a different place than the current drive and directory.

When saving a file, STEPS always looks on the disk to see if the file already exists. If it does, STEPS renames the old file with a BAK extension before your file is saved on the disk.

Printing STEPS Programs

One of the crucial parts of Ladder Logic programming is good documentation. It is extremely important to be able to review programs at any time in appropriate formats. STEPS offers three different ways to print programs so they can be used as working documents, or for reference purposes.

To print your program, first make sure your printer is on-line and you are looking at the STEPS main menu. STEPS always prints using the LPT1 port on your computer. Press the P key to access the Printing menu and select the type of listing desired. Unless you choose the Text Only selection, STEPS will instruct you to enter IN if your printer requires nongraphic characters. Most printers used with personal computers support the graphics characters that are used by STEPS to draw logic diagrams on the screen. However, there are still many printers in use that do not—in which case you should press N and then the Enter key. If your printer does support graphics characters, then just press Enter. If you are not sure, then you can try it both ways and find out for sure.

Note: Laserjet II users, select PC-8 character set to get graphics for logic diagrams.

The most complete format offered includes text and ladder logic combined in one listing. This format prints each line of the program file as text, but also precedes all Logic Lines with a logic diagram just as it would appear on the screen.

The Ladder Only Format prints only Logic Lines in ladder format with sides and all. This is the format that most accurately resembles an electrical relay logic drawing. Also, this type of listing presents the most information in a brief format.

The Text Only Format is precisely that—a listing of the text file, which is a STEPS program. If you don't mind crowding all your comments and Logic Lines into one place, this can be combined with the Ladder Only listing (either before or after) to make a complete document.

Transferring Programs to The StepLadder PLC

Before you attempt to transfer your program, make sure your personal computer is properly hooked up to The StepLadder PLC. The StepLadder must be plugged into an IO module rack with a five-volt power supply and the power must be turned ON. To initiate the transfer, press T at the STEPS Main menu. You will be presented with a communications menu that allows you to change the COM port or data rate to suit the StepLadder that you are connected to.

Then STEPS will go through your program to translate it to the form that The StepLadder PLC requires. You will be able to watch this process as each line of the program is displayed on the screen. If a problem is found, the line containing the problem is shown along with a message. Then the transfer process is terminated, so that you can edit your program.

When STEPS has completed its translation, there will be a brief delay and then some characters will appear on the screen followed by a long stream of backslash ("\") characters. This means that the actual transfer is underway. When complete, STEPS will ask if you want to restart the StepLadder. This gives you the opportunity to check the conditions of your process before actually executing the new or revised logic. You can type Y and Enter whenever you are ready, or you can type N and Enter to make STEPS return to its main menu. If you elect not to restart the StepLadder, it will start the next time it is powered up.

Communicating with The StepLadder PLC

The StepLadder PLC has communications features which can be adapted to a wide variety of applications. These features can be used to get information from the StepLadder, or to deliver information to the StepLadder which can then be used in the ladder logic program. The status of IO points, internal relays, and current timer or counter information are all accessible via RS232, point-to-point communications.

You can also connect two StepLadders together and have them share IO point information. Each unit can be programmed to act on data from the other. Thus, the two StepLadders can become a forty eight point PLC system.

The StepLadder PLC serial port

The StepLadder's serial port is RS232 compatible and can be used at 1200, 2400, 4800 or 9600 baud with 8 data bits, no parity, and 1 stop bit. See the section on *StepLadder connections* for wiring details.

Caution

When communicating with The StepLadder PLC in electrically noisy environments or over long distances, it is advisable to convert the RS232 serial data to another type of electrical signal. Many devices are available for this purpose, from short-haul modems to fiberoptic devices.

Communications Initiated by a Host

A Host computer, or any ASCII terminal, can be used to communicate with The StepLadder PLC. A command-and-response method is used for this type of communication. The host will issue a command, and the StepLadder will take action on that command. The action may be an operation at one of the IO points, or, if the command is a data request, the StepLadder will respond with the appropriate data.

Throughout this section the commands are described as if the host were a terminal or a Personal Computer running terminal emulation software. The StepLadder PLC will issue a prompt in response to a carriage return character (or the Enter key on the PC keyboard). This prompt tells you that you are connected to a StepLadder and gives you its version number.

Example   StpL v2.0>

Hexadecimal response

The carriage return character also sets the response mode to Hexadecimal in the StepLadder. This means that data sent from the StepLadder will be in hexadecimal characters 0 through 9 and A through F. Each byte of data will be sent as two hexadecimal characters. Each word of data, such as 16-bit timer and counter values, will be sent as four hexadecimal characters. The first character sent is always the most significant.

Binary response

You can select binary response by sending a period (".") to the StepLadder. A short prompt (">") will be issued to acknowledge binary response. In this mode, the commands are the same but the StepLadder's response is in binary instead of hexadecimal form. Also, no spaces or carriage returns are sent, which makes the data hard to see on a terminal. However, there are speed advantages to using this feature if you want to write software to communicate with the StepLadder; you can get data from the StepLadder in less than half the time using binary response.

Data Request Commands

To request data from The StepLadder PLC, simply enter the appropriate lowercase character (a through e). The StepLadder will respond with a line of hexadecimal characters when in hexadecimal response mode; the same data will come out in pure binary, with no spaces or carriage return, if in binary response mode.

Each byte (or 2 hex characters) represents eight IO points or internal relays. The most significant bit in the byte represents the first of the eight points. For instance, the first eight IO points sent as hex 80 would mean IO point zero is ON. The first eight IO points sent as hex 01 would mean IO point seven is ON. This allows you to see the hexadecimal characters in the same order as the discrete points appear on the IO rack. When looking at an IO rack, you can see point zero in the left-most position in most cases.

Timers are shown as 16-bit counts representing the current time in $\frac{1}{100}$ of a second. Counters are also shown as 16-bit counts but representing the current count. The scan count is an 8-bit count incremented at the completion of each scan. The check sum is the sum of all previous data. The data included in the five available data requests is as follows.

The first data request is called the General Data Request 'a.

```
'a
        IO points                0   -    7
        IO points                8   -   15
        IO points               16   -   23
        Internal Relays         96   -  103
        Internal Relays        104   -  111
        Internal Relays        112   -  119
        Internal Relays        120   -  127
        Internal Relays        128   -  135
        Internal Relays        136   -  143
        Timer 0
        Timer 1
        Scan Count
        Check Sum
```

Example   >a80 3B 7E 04 C0 02 00 00 00 0001 0002 D8 DA
               >

Data requests 'b, 'c, and 'd provide current counter information.

'b
Counter 0
Counter 1
Counter 2
Counter 3
Check Sum

'c
Counter 4
Counter 5
Counter 6
Counter 7
Check Sum

'd
Counter 8
Counter 9
unused
unused
Check Sum

The 'e data request includes the remaining internal relays not including in the General data request.

'e

```
Internal Relays    144   - 151
Internal Relays    152   - 159
Internal Relays    160   - 167
Internal Relays    168   - 175
Internal Relays    176   - 183
Check Sum
```

The following is an example of the check sum calculation. A b has been sent to The StepLadder PLC to request counter data.

Example   >b0005 0102 0300 0001 0C

>

00 + 05 + 01 + 02 + 03 + 00 + 00 + 01 = 0C

Point Operate Command

Physical IO points and internal relays can be operated from a host with the point operate command. All point operate features can be used to operate internal relays. There are some restrictions on which of the point operate features can be used with physical IO points.

Physical IO points must be designated as outputs in the StepLadders program before they can be used in a point operate command. Only the Latch and Reset op codes can be used on physical IO points. These limitations can easily be dealt with by directing the Point Operate Command to an internal relay. Then simply operate the Physical Point with a logic line that uses the internal relay in its logic definition.

Four types of operations (op codes) are available. Set and Clear turn the point ON or OFF. Latch turns the point ON and latches it so that it can only be cleared by a Reset, which clears the latch and turns the point OFF. Points can be set and cleared or latched and reset by logic in the StepLadders program. So one must be aware of the program in operation when using this feature.

The format of the point operate command is:

```
Pxx/y"
```

where   P   Designates the point operate command.

        xx   The point number in hexadecimal.

        /   Designates the op code to follow.

        y   The opcode (0,1,2 or 3)

        "   Executes the command

Op codes are

0   CLEAR, turns the point OFF

1 SET, turns the point ON
2 LATCH, turns ON the point and latches it.
3 RESET, turns OFF the point and its latch.

Example   >P60/2"OK

This command latches internal relay 96 (hex 60). The StepLadder will respond with 'OK to indicate that the Point Operate command has been executed. See Part C in this appendix for error codes.

The Store Command

Sometimes it becomes convenient for the host to have the ability to change sixteen bits of data in The StepLadder PLC. A counter preset or the actual count are examples of this kind of data. Also, it is possible to set sixteen internal relay states with a single Store Command.

Each piece of 16-bit data that can be changed with this command has been assigned a register number. The format of the Store Command is:

xxxx/yy!

where

xxxx   The data to be stored in hex
/   Designates the register number follows
yy   The register number
!   Executes the command

Example   >000C/1A!OK

changes the preset for counter 8 to an even dozen (C). Notice that the StepLadder responds with "OK." See Part C at the end of this appendix for error codes.

Valid register numbers are

```
00      Internal relays 96 through 111
01      Internal relays 112 through 127
02      Internal relays 128 through 143
03      Internal relays 144 through 159
04      Internal relays 160 through 175

05      Tmr 0 time           06      Tmr 0 preset
07      Tmr 1 time           08      Tmr 1 preset
09      Cntr 0 count         0A      Cntr 0 preset
0B      Cntr 1 count         0C      Cntr 1 preset
0D      Cntr 2 count         0E      Cntr 2 preset
0F      Cntr 3 count         10      Cntr 3 preset
11      Cntr 4 count         12      Cntr 4 preset
13      Cntr 5 count         14      Cntr 5 preset
```

15	Cntr 6 count	16	Cntr 6 preset
17	Cntr 7 count	18	Cntr 7 preset
19	Cntr 8 count	1A	Cntr 8 preset
1B	Cntr 9 count	1C	Cntr 9 preset

As with other discrete data, the lowest number internal relay in each register is in the left-most position.

Example   >8000/00!OK

Sets point number 96 (first internal relay) ON. To set points 96 and 111 ON enter the following.

Example   >8001/00!OK

Commands to Control The StepLadder PLC

Sometimes it is necessary to restart The StepLadder PLC from a host. There are commands available for this purpose.

The HALT Command stops the scanning process. All IO points and internal relays are left in their previous states. While halted the StepLadder cannot alter any output points. The format of the halt command is four uppercase H characters ("HHHH"). These characters must be received by the StepLadder within approximately two and a half seconds in order to accomplish their purpose. Otherwise, they will be ignored. A prompt character (">") will be issued by the StepLadder (possibly two prompts) to acknowledge that it has halted. You can check to see if the StepLadder is halted with a single H character. The prompt will only be issued when halted.

Example >HHHH>

The GO command simply resumes the scanning process when it has been halted. If the StepLadder has not been halted this command does nothing. A single upper-case G is all that is required.

Example   >G

The RESET command accomplishes the same thing as a power up of the StepLadder. All IO points and internal relays are cleared, the nonvolatile memory, containing the logic program is checked, and all timer and counter presets are set to their programmed values. Then the scanning process begins. This command is not accepted unless the StepLadder is halted. The format is simply an uppercase R character. When this command is issued, The StepLadder PLC immediately initializes itself, including its communications hardware. Therefore the R character is not echoed like many other commands and does not appear at the host.

Communications Initiated by The StepLadder PLC

Sometimes it becomes appropriate for the StepLadder to send information indicating that a certain event has occurred. Thus, a host system doesn't have to poll the

StepLadder when it is waiting for such an event.

The StepLadder can be programmed to initiate either of two types of communications. One is called the Event Flag, which is simply a single character ("#") sent out to indicate that the programmed event has occurred. The other is called a Block Transfer. This is a block of data that is sent out as a result of the event.

Event Flag # (IO point 188)

To use the Event Flag feature, simply use the Event Flag IO point (188) as the output of a logic line in your program. In most cases, it should be designated as a One-Shot type output so that only one number sign character ("#") is sent for each event. Otherwise a stream of number sign characters will be generated as long as the result of the logic line is true.

Note: If you use the output of a timer to operate the event flag (#188) or the Block transfer (#187), two events will occur. Therefore it is usually better to assign timer outputs to an internal relay and then use that point to operate the event flag or block transfer point numbers.

Block Transfer (IO Point 187)

To use the Block Transfer feature, simply use the Block Transfer IO point (187) as the output of a logic line in your program. As with the Event Flag, in most cases it should be designated as a One-Shot type output so that only one Block Transfer is sent for each event.

The Block Transfer is seven bytes long. The first byte is the dollar sign character ("$"). This is followed by the body; a check sum byte, and the number sign character ("#"), which indicates the end of the Block Transfer. The body of the Block Transfer has four bytes (32 bits) of actual data, which includes the twenty four physical IO points and eight of the internal relays.

```
byte 1        '$', Start of Block
byte 2        IO points 0 through 7          body
byte 3        IO points 8 through 15         body
byte 4        IO points 16 through 23        body
byte 5        Internal relays 176 through 183 body
byte 6        Check sum of 4 previous bytes
byte 7        '#', End of Block
```

The StepLadder PLC is also capable of receiving a Block Transfer. This is the mechanism used to share data between two StepLadders and thus create a larger PLC system. When a valid Block Transfer arrives, the four bytes of data (32 bits) are assigned to internal relays 144 through 175 in the receiving StepLadder. So internal relay 144 at the receiving unit becomes the same state as IO point 0 of the unit that sent the Block Transfer. Both StepLadders can send and receive Block Transfers at the same time. Thus, both units have access to the other's IO point data.

The table here shows the originating point numbers at a StepLadder that is sending a Block Transfer, and the point numbers where the receiving StepLadder sees the same data.

SND	RCV	SND	RCV	SND	RCV	SND	RCV
0 –	144	8 –	152	16 –	160	176–	168
1 –	145	9 –	153	17 –	161	177–	169
2 –	146	10 –	154	18 –	162	178–	170
3 –	147	11 –	155	19 –	163	179–	171
4 –	148	12 –	156	20 –	164	180–	172
5 –	149	13 –	157	21 –	165	181–	173
6 –	150	14 –	158	22 –	166	182–	174
7 –	151	15 –	159	23 –	167	183–	175

Part A: StepLadder Connections

The communications connector on The StepLadder PLC is a six position RJ (telephone) type connector. It provides the RS232 interface for connection to a personal computer or serial device. Pin 1 is located closest to the edge of the board and is identified on the bottom of the circuit board. The signals on the connector are as follows:

pin	signal
1	No connection
2	Input (not currently used)
3	Serial data into StepLadder
4	Serial data out of StepLadder
5	Signal ground
6	No connection

The 50-pin interface connector is designed to mate directly with IO module racks that use header type connectors for interface. All even-numbered pins (2–50) are connected to ground and are labeled on the top of the board with 50 and 2 at each end of the row. Odd numbered pins are assigned to IO points with the exception of pin 49, which is used for five-volt power to The StepLadder PLC.

pin	signal	pin	signal
1	IO point 23	25	IO point 11
3	IO point 22	27	IO point 10
5	IO point 21	29	IO point 9
7	IO point 20	31	IO point 8
9	IO point 19	33	IO point 7
11	IO point 18	35	IO point 6
13	IO point 17	37	IO point 5
15	IO point 16	39	IO point 4

```
17      IO point 15        41      IO point 3
19      IO point 14        43      IO point 2
21      IO point 13        45      IO point 1
23      IO point 12        47      IO point 0

pin 49                     + 5 volts DC
even pins 2-50             power supply common
```

IO points 0–23 are designed to interface with IO modules that use five-volt negative true logic.

Part B: Hexadecimal Point Numbers

This table can be used to find hexadecimal designations for all IO point numbers. Also, the corresponding register numbers are included for Internal Relays. Hex characters are shown as bold in the table.

Physical IO point numbers

0	1	2	3	4	5	6	7
00	**01**	**02**	**03**	**04**	**05**	**06**	**07**

8	9	10	11	12	13	14	15
08	**09**	**0A**	**0B**	**0C**	**0D**	**0E**	**0F**

16	17	18	19	20	21	22	23
10	**11**	**12**	**13**	**14**	**15**	**16**	**17**

Internal relay point numbers

96	97	98	99	100	101	102	103	00
60	**61**	**62**	**63**	**64**	**65**	**66**	**67**	ms

104	105	106	107	108	109	110	111	00
68	**69**	**6A**	**6B**	**6C**	**6D**	**6E**	**6F**	1s

112	113	114	115	116	117	118	119	01
70	**71**	**72**	**73**	**74**	**75**	**76**	**77**	ms

120	121	122	123	124	125	126	127	01
78	**79**	**7A**	**7B**	**7C**	**7D**	**7E**	**7F**	1s

128	129	130	131	132	133	134	135	02
80	**81**	**82**	**83**	**84**	**85**	**86**	**87**	ms

| 136 | 137 | 138 | 139 | 140 | 141 | 142 | 143 | 02 |
| 88 | 89 | 8A | 8B | 8C | 8D | 8E | 8F | 1s |

| 144 | 145 | 146 | 147 | 148 | 149 | 150 | 151 | 03 |
| 90 | 91 | 92 | 93 | 94 | 95 | 96 | 97 | ms |

| 151 | 153 | 154 | 155 | 156 | 157 | 158 | 159 | 03 |
| 98 | 99 | 9A | 9B | 9C | 9D | 9E | 9F | 1s |

| 160 | 161 | 162 | 163 | 164 | 165 | 166 | 167 | 04 |
| A0 | A1 | A2 | A3 | A4 | A5 | A6 | A7 | ms |

| 168 | 169 | 170 | 171 | 172 | 173 | 174 | 175 | 04 |
| A8 | A9 | AA | AB | AC | AD | AE | AF | 1s |

| 176 | 177 | 178 | 179 | 180 | 181 | 182 | 183 |
| B0 | B1 | B2 | B3 | B4 | B5 | B6 | B7 |

Part C: Error Codes Returned to Host

The Point Operate and Store commands, both return OK after they have received and executed a valid command. However, when the command is not understood by The StepLadder PLC, an error code will be sent to the host. Error codes are two characters long with no control codes or white space characters (CR, LF etc.)

1? Quote character was received without proper point operate command.
2? Slash missing from point operate command.
3? Invalid point number.
4? Invalid opcode for point operate command.
5? Not used.
6? Slash missing from Store command.
7? Invalid register number.
8? Not used.
9? Receive buffer overflow.

Appendix D

Boolean Algebra Summary

Operations

$\times$ = AND

$+$ = OR

$\overline{A}$ = NOT A

Laws

Commutative

$A+B = B+A$

$AB = BA$

Associative

$A+(B+C) = (A+B)+C$

$A(BC) = (AB)C$

Distributive

$A(B+C) = AB+AC$

$A+BC = (A+B)\ (A+C)$

Absorptive

$A(A+B) = A$

DeMorgan's

$\overline{(A+B+C)} = \overline{A}\overline{B}\overline{C}$

$\overline{(ABC)} = \overline{A}+\overline{B}+\overline{C}$

Identities

$AA = A$

$X(1+A+B+C+\ldots) = X$

$\overline{\overline{A}} = A$

$\overline{\overline{(AB)}} = AB$

$\overline{\overline{(A+B)}} = A+B$

$AB+A\overline{B} = A$

$A+\overline{A}B = A+B$

$AB+AC+B\overline{C} = AC+B\overline{C}$

ASCII Character Code

Binary	Octal	Decimal	Hexadecimal	ASCII	Remarks
0000000	000	000	00	NUL	Null
					Tape leader
					Ctrl-shift P
0000001	001	001	01	SOH	Start of heading
				SOM	Start of message
					Ctrl-A
0000010	002	002	02	STX	Start of text
				EOA	End of address
					Ctrl-B
0000011	003	003	03	ETX	End of text
				EOM	End of message
					Ctrl-C
0000100	004	004	04	EOT	End of transmission
					End
					Ctrl-D
0000101	005	005	05	ENQ	Enquiry
					Who are you
					Ctrl-E
0000110	006	006	06	ACK	Acknowledge
					Are you
					Ctrl-F
0000111	007	007	07	BEL	Ring bell
					Ctrl-G
0001000	010	008	08	BS	Backspace
					Form effector
					Ctrl-H

Binary	Octal	Decimal	Hexadecimal	ASCII	Remarks
0001001	011	009	09	HT	Horizontal tab
					Tab
					Ctrl-I
0001010	012	010	0A	LF	Line feed
					New line
					Ctrl-J
0001011	013	011	0B	VT	Vertical tab
					Ctrl-K
0001100	014	012	0C	FF	Form feed
					Page
					Ctrl-L
0001101	015	013	0D	CR	Carriage return
					End of line
					Ctrl-M
0001110	016	014	0E	SO	Shift out
					Red-colored ribbon
					Ctrl-N
0001111	017	015	0F	SI	Shift-in
					Black-colored ribbon
					Ctrl-O
0010000	020	016	10	DLE	Data link escape
					Ctrl-P
0010001	021	017	11	DC1	Device control 1
					Transmitter on
					Reader on
					Ctrl-Q
0010010	022	018	12	DC2	Device control 2
					Tape punch on
					Ctrl-R
0010011	023	019	13	DC3	Device control 3
					Transmitter off
					Reader off
					Ctrl-S
0010100	024	020	14	DC4	Device control 4
					Tape punch off
					Ctrl-T
0010101	025	021	15	NAK	Negative acknowledge
					Error
					Ctrl-U
0010110	026	022	16	SYN	Synchronous file
					Synchronous idle
					Ctrl-V
0010111	027	023	17	ETB	End of text buffer
					Logical end medium
					Ctrl-W

0011000	030	024	18	CAN	Cancel Ctrl-X
0011001	031	025	19	EM	End of medium Ctrl-Y
0011010	032	026	1A	SUB	Substitute Ctrl-Z
0011011	033	027	1B	ESC	Esc Prefix Ctrl-shift K
0011100	034	028	1C	FS	File separator Ctrl-shift L
0011101	035	029	1D	GS	Group separator Ctrl-shift M
0011110	036	030	1E	RS	Record separator Ctrl-shift N
0011111	037	031	1F	US	Unit separator Ctrl-shift O
0100000	040	032	20	SP	Space Blank
0100001	041	033	21	!	
0100010	042	034	22	"	
0100011	043	035	23	#	
0100100	044	036	24	$	
0100101	045	037	25	%	
0100110	046	038	26	&	
0100111	047	039	27	'	Apostrophe
0101000	050	040	28	(	
0101001	051	041	29	)	
0101010	052	042	2A	*	
0101011	053	043	2B	+	
0101100	054	044	2C	,	Comma
0101101	055	045	2D	-	
0101110	056	046	2E	.	Period
0101111	057	047	2F	/	
0110000	060	048	30	0	Number 0
0110001	061	049	31	1	Number 1
0110010	062	050	32	2	
0110011	063	051	33	3	
0110100	064	052	34	4	
0110101	065	053	35	5	
0110110	066	054	36	6	
0110111	067	055	37	7	
0111000	070	056	38	8	
0111001	071	057	39	9	
0111010	072	058	3A	:	
0111011	073	059	3B	;	

Binary	Octal	Decimal	Hexadecimal	ASCII	Remarks
0111100	074	060	3C	<	
0111101	075	061	3D	=	
0111110	076	062	3E	>	
0111111	077	063	3F	?	
1000000	100	064	40	@	
1000001	101	065	41	A	
1000010	102	066	42	B	
1000011	103	067	43	C	
1000100	104	068	44	D	
1000101	105	069	45	E	
1000110	106	070	46	F	
1000111	107	071	47	G	
1001000	110	072	48	H	
1001001	111	073	49	I	Letter I
1001010	112	074	4A	J	
1001011	113	075	4B	K	
1001100	114	076	4C	L	
1001101	115	077	4D	M	
1001110	116	078	4E	N	
1001111	117	079	4F	O	Letter O
1010000	120	080	50	P	
1010001	121	081	51	Q	
1010010	122	082	52	R	
1010011	123	083	53	S	
1010100	124	084	54	T	
1010101	125	085	55	U	
1010110	126	086	56	V	
1010111	127	087	57	W	
1011000	130	088	58	X	
1011001	131	089	59	Y	
1011010	132	090	5A	Z	
1011011	133	091	5B	[	
1011100	134	092	5C	\	
1011101	135	093	5D	]	
1011110	136	094	5E	^	Circumflex
1011111	137	095	5F	_	Underline
1100000	140	096	60	`	Accent grave
1100001	141	097	61	a	
1100010	142	098	62	b	
1100011	143	099	63	c	
1100100	144	100	64	d	
1100101	145	101	65	e	
1100110	146	102	66	f	
1100111	147	103	67	g	
1101000	150	104	68	h	

1101001	151	105	69	i	
1101010	152	106	6A	j	
1101011	153	107	6B	k	
1101100	154	108	6C	l	Lowercase l
1101101	155	109	6D	m	
1101110	156	110	6E	n	
1101111	157	111	6F	o	
1110000	160	112	70	p	
1110001	161	113	71	q	
1110010	162	114	72	r	
1110011	163	115	73	s	
1110100	164	116	74	t	
1110101	165	117	75	u	
1110110	166	118	76	v	
1110111	167	119	77	w	
1111000	170	120	78	x	
1111001	171	121	79	y	
1111010	172	122	7A	z	
1111011	173	123	7B	{	
1111100	174	124	7C	|	Vertical bar
1111101	175	125	7D	}	
1111110	176	126	7E	~	Tilde
1111111	177	127	7F	Del	Del
					Rubout

Note that binary codes 0000000 through 0011111 are control codes and are thus not printed. Binary codes 0100000 through 1111111 actually produce printed characters.

Summary of Relay Ladder & Boolean Programming Symbols

Ladder symbols	Definition
—‖—	N.O.
—‖̸—	N.C.
—( )—	Out
—(/)—	Out not
—(L)—	Latch out
—(U)—	Unlatch out
—(TIM)—	Timer
—(CNT)—	Counter

Ladder symbols	Definition
—(+)—	Addition
—(–)—	Subtraction
—(x)—	Multiplication
—(÷)—	Division
101 —| GET |—	Get instruction
102 —(PUT)—	Put instruction
—| ↑ |—	Off-to-on transitional
—| ↓ |—	On-to-off transitional

} one-shot contacts

| —(MCR)— | Master control relay instruction |
| —(END MCR)— | End MCR instruction |

Ladder symbols	Definition

—(ZCL)—   Zone control last instruction

—(END ZCL)—   End ZCL instruction

103
—(JMP)—   JUMP instruction

104
—(JSB)—   Jump-to-subroutine instruction

—(RET)—   Return-from-subroutine instruction

105
—|CMP=|—   Compare equal

106
—|CMP>|—   Compare greater than

107
—|CMP<|—   Compare less than

Boolean Symbols with Ladder Equivalents

Ladder diagram	Boolean mnemonic
—| |—	AND
—| |— (with branch)	OR
—( )—	OUT
—(/)—	OUT NOT
—|/|—	NAND
—|/|— (with branch)	NOR
—| |—	LOAD
—| /|—	LOAD NOT
—(L)—	OUT L
—(U)—	OUT U
—(TIM)—	TIM
—(CNT)—	CNT

Ladder diagram	Boolean mnemonic
—(+)—	ADD
—(–)—	SUB
—(x)—	MUL
—(÷)—	DIV
—(CMP =)—	CMP =
—(CMP >)—	CMP >
—(CMP <)—	CMP <
—(JMP)—	JMP
—(JSB)—	JSB
—(MCR)—	MCR
—(END MCR)—	END

G

Selected Supplier Literature

This appendix contains technical literature on a variety of PCs: small, medium, and large. Included is information on an interface module that allows access to a local area network. Readers who have completed the text of this book will find this appendix very informative. It presents the opportunity to test their understanding of the material covered in the text.

Inclusion of suppliers' literature in the appendix does not constitute an endorsement of the suppliers' products.

Literature on Westinghouse products is reprinted with the permission of Westinghouse Electric Corporation. Literature on the Director 6001 PC is reprinted with permission of UTICOR Technology, Inc. Literature on the Eagle Series of PCs is reprinted with the permission of Eagle Signal Controls, a Division of Mark IV Industries, Inc. Literature on the SY/NET Transfer Network Interface Module is reprinted with the permission of Square D Company.

Small systems

Reprinted with permission of Westinghouse Electric Corporation

Westinghouse 50 Series

Introduction

The Westinghouse 50 Series Programmable Logic Controllers (PLCs) combine the CPU power, analog processing, I/O flexibility and networking capabilities of mid-range controllers with the benefits of small, fixed I/O PLCs. The PC-50 and PC-55 are compact, fixed I/O PLCs that can be expanded using the full complement of Westinghouse 500 Series input/output modules.

The 50 Series PLCs provide the user, or OEM, unmatched performance through the unique combination of powerful, high-speed CPUs, on-board I/O, and complete program compatibility with the entire Westinghouse PLC family.

Designed specifically for applications that require power and speed of a full-function PLC without the need for a large amount of I/O, the PC-50 is well suited for applications including:
— Machine control
— Compressor control
— Pump control
— Bottling and filling lines

The PC-55 has complete analog processing capability, real-time clock (future), floating point math, and a complement of analog, high speed counting, and digital I/O onboard with the CPU and power supply. This combination provides unique cost and space effective solutions for many applications, including:
— Packaging machinery
— Heating and Air-conditioning control systems
— Material handling
— Pump station control
— Dairy CIP Control

Westinghouse 50 Series Features

Both 50 Series programmable controllers, the PC-50 and the PC-55, consist of a base unit which includes power supply, CPU and I/O.

PC-50 Features include:
- 8 digital inputs (24 VDC), 1 interrupt input (25 VDC), 1 high speed counter input and 6 relay outputs in less than 34 square inches of panel space.

- The CPU can be powered directly from a 115/230 VAC power source.

- 2K words of battery backed RAM memory for program storage. Optional EPROM or EEPROM cartridges for long-term memory security.

- I/O expansion with up to six 500 Series I/O modules for the flexibility to add the digital, analog or special purpose I/O needed to meet a particular application.

PC-55 features include:
- 16 digital inputs, 16 digital outputs, 8 analog inputs, 1 analog output, 4 interrupt inputs and 2 high speed counter inputs in a 45 square inch footprint.

- 4 interrupt inputs suspend normal CPU logic processing for fast response to alarm conditions.

- 2 high speed counter inputs for positioning and totalizing applications.

- The analog and interrupt inputs can optionally be used as digital inputs. This provides up to 28 digital inputs (24 VDC) with no additional hardware.

- 8K words of RAM memory (4K for program and 4K for data storage). Program back-up is with battery or optional EPROM and EEPROM cartridges.

- I/O expansion with up to thirty-two 500 Series I/O modules. Choose the combination of digital, analog or special purpose I/O you need.

- Advanced processor features include PID control, floating point math capability and a real-time clock (future). Perfect for sequencing and control of pumps, compressors and HVAC systems.

Base Units Overview

PC-50 Base Unit

The PC-50 operates directly from a 115 or 230 VAC power source connected to screw terminals located on the top front of the base unit. A Run/Stop switch, along with LEDs indicating the operating mode of the CPU and the status of each individual channel, are also conveniently located on the face of the unit.

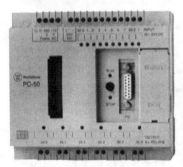

Logic functions provided by the PC-50 include set/reset, timers, counters, and comparators. The 2K words of internal RAM memory can be backed-up by an optional battery. EPROM and EEPROM memory submodules for permanent, non-volatile program storage are also available.

PC-50 Onboard I/O

The PC-50 base unit includes I/O circuitry to handle the following inputs/outputs:

- 8 – 24 VDC inputs
- 1 – 24 VDC interrupt input
- 1 – 24 VDC high speed counter input
- 6 – Relay contact outputs

Field wiring connection to these input/outputs are made by screw terminals at the top (inputs) and bottom (outputs) front of the base unit.

The PC-50 Base Unit can be mounted either to a standard Din rail, or directly to the panel using optional snap-in mounting brackets.

The Input/Output capacity of the PC-50 can be expanded with up to six 500 Series modules. This will provide a total of up to 48 digital/8 analog signals. Connection to the 500 Series bus modules is made via a DIN-rail mounted interface module (future).

Programming of the PC-50 is done via the Programming/L1 network port located on the face of the PC-50 Base Unit. The PC-50 can be programmed using either a handheld loader or personal computer software loader. A full-function version of the program loader software, NLSW-3103, is available for use solely with the PC-50. The PC-50 can also be programmed with either NLSW-3104 or NLSW-3105.

PC-55 Base Unit

The PC-55 is designed to operate from a 24VDC power source. If 115V or 230V AC is to be used, an optional power supply module is available.

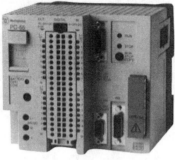

All standard logic functions, as well as advanced functions including analog scaling, indirect addressing, PID loop control algorithm, and a realtime clock (future) are included in the PC-55 function set. 4K words of RAM memory is available for the user logic program with an additional 4K words of memory reserved for data storage. The logic program can be protected during power outages using an optional battery. Permanent memory protection is available with a plug-in EPROM or EEPROM memory cartridge.

PC-55 Onboard I/O

The PC-55 base unit includes I/O circuitry to handle the following inputs/outputs:

- 16 – 24 VDC inputs
- 16 – 24 VDC outputs
- 4 – 24 VDC interrupt inputs
- 2 – High speed counter inputs
- 8 – Analog inputs (0-10V)
- 1 – Analog output (0-10V or 4-20ma)

The 32 digital inputs/outputs are terminated to a 40-pin removable terminal block NLTB-500 located on the face of the PC-55. The NLTB-500 must be purchased separately. The interrupt, counter and analog signal termination is via two "D" connectors also located on the face of the base unit. DIN rail mounted screw terminal blocks/cable assemblies are available to provide a convenient screw terminal interface.

The PC-55 Base Unit can be mounted either to a standard DIN rail, or directly to the panel using optional snap-in mounting brackets.

The PC-55 can be expanded using the full complement of 500 Series I/O modules. A total of 32 modules can be added to the PC-55 to provide up to the maximum 256 digital/16 analog I/O available. An interface module is not required to connect 500 Series I/O modules to the PC-55.

Either a handheld loader or PC software (NLSW-3104, NLSW-3105) can be used to program the PC-55.

Technical Specifications

Catalog Number Feature	PC-50	PC-55
● Internal RAM memory	2K Words	4K Words Logic and 4K Words Data
● Approximate scan time	2msec/K	2msec/K
● Supply Voltage	115VAC (93-127V) 230VAC (187-253V)	24VDC (20-30VDC)
● Flags – Total –Retentive	1024 512	2048 512
● Timers – Quantity – Time range	32 0.01 to 9990 sec	128 0.01 to 9990 sec

● Counters – Total
  – Retentive
  – Range
	32
	8
	0 to 999
		128
		8
		0 to 999

● Data Blocks	2..63	2...255
● Program Blocks	0..63	0..255

● Function Blocks
  Programmable
  Integral
	
	0..63
	None
		
		0..255
		BCD-Binary, Binary –
		BCD, Multiply,
		Divide
		Analog Scaling

● Organizational Blocks
  OB1 Cyclic Scan
  OB3 Interrupt Scan
  OB13 Time-Controlled
  OB21 Scan on Run
  OB22 Scan on Power Up
  OB31 Reset Watchdog
  OB34 Battery Fail
  OB251 PID Algorithm
	
	Yes
	Yes
	No
	Yes
	Yes
	No
	No
	No
		
		Yes
		Yes
		Yes
		Yes
		Yes
		Yes
		Yes
		Yes

● Real Time Clock	No	Yes (future)

● Dimensions (W x H x D)
  inches
  mm
	
	5.7 x 5.3 x 3.6
	145 x 135 x 91
		
		5.7 x 5.3 x 5.5
		145 x 135 x 146

● Program Loader Software	NLSW-3103 NLSW-3104 NLSW-3105	NLSW-3104 NLSW-3105
● Battery Back-up	NLB-50	NLB-500
● EPROM Modules (K = 1,000 Words)	NLEP-50 (2K)	NLEP-2508 (4K) NLEP-2516 (8K)
● EEPROM Modules (K = 1,000 Words)	NLEE-50 (2K)	NLEE-2502 (1K) NLEE-2504 (2K) NLEE-2508 (4K) NLEE-2516 (8K)

PC-50 and PC-55 common features include:
- Scan speeds of less than 2 msec/K of logic allow the 50 Series to easily meet the demanding requirements of packaging machines, material handling systems, robotics, and machine tools.

- Two programming languages, Ladder Diagram for straight forward logic tasks, or Statement List for more complex applications provide the programmer with the most efficient language for the specific task.

- Structured programming permits more efficient program development and faster throughput by executing only the portions of the program actually needed; allows repeated use of similar functions without added programming.

- Three versions of program loaders:

  Software for IBM™ Personal Computer or compatible. On-line programming and monitoring, off-line programming and full documentation capability in both ladder diagram and statement list modes.

  Hand held with 8-line LCD display. On-line programming and monitoring, off-line programming in statement list mode. EPROM and EEPROM programming.

  Hand held with 2-line LCD display. On-line programming and monitoring in statement list mode.

- Mounts on a standard DIN rail, or directly to panel with snap-in adapters.

- Components are UL listed, CSA certified and meet applicable IEC and UDE standards.

- All Westinghouse PLC components carry a full one-year warranty on material and workmanship.

Onboard I/O Technical Specifications

Onboard I/O	PC-50
Digital Inputs	
Number of inputs	8
Inputs per common	All
Input voltage	
Rated	24 V DC
OFF voltage level	0 to 5 V DC
ON voltage level	13 to 30 V DC
Input loading	8.5 ma
Response time	
OFF to ON (typ.)	2.8 ms
ON to OFF (typ.)	3.6 ms
Interrupt Inputs	
Number of interrupt inputs	1
Response time	
OFF to ON (typ.)	40 μs
ON to OFF (typ.)	180 μs
Min. pulse duration	500 μs
Counter Inputs	
Number of counters	1
Counter frequency	1 kHz
Response time	
OFF to ON (typ.)	40 μs
ON to OFF (typ.)	180 μs
Min. pulse duration	500 μs
Relay Outputs	
Number of relay outputs	6
Isolated in groups of	1
Contact Rating	
Resistive load (max.)	3A at 250 V AC
	1.5A at 30 V DC
Inductive load (max.)	0.5A at 250 V AC
	0.5A at 30 D DC

Onboard I/O	PC-55
Digital Inputs	
Number of inputs	16
Inputs per common	16
Input voltage	
Rated	24 V DC
OFF voltage level	− 30 to + 5 V DC
ON voltage level	+ 13 to + 30 V DC
Input loading	<6.5ma
Response time	
OFF to ON (typ.)	2.5 ms
ON to OFF (typ.)	2.0 ms
Interrupt Inputs	
Number of interrupt inputs	4
Response time	
OFF to ON (typ.)	75 μs
ON to OFF (typ.)	140 μs
Min. pulse duration	500 μs
Counter Inputs	
Number of counters	2
Counter frequency	Counter A 5 kHz
	Counter B 2 kHz
Response time	
OFF to ON (typ.)	10μs
ON to OFF (typ.)	15 μs
Min. pulse duration	100 μs
Digital Outputs	
Number of digital outputs	16
Isolated in groups of	16
Load Voltage	
Rated	24 V DC
Permissable range	20 to 30 V DC
Output current (ON)	
Per circuit	max. 0.5A
All circuits	6A
	8A (<50°C)
Leakage current	<50 μA
Heat dissipation	max. 50W
Analog Inputs	
Number of analog inputs	8
Input range	0 to + 10 V
Resolution	11 Bits
Input impedance	20 kΩ
Analog Outputs	
Number of analog outputs	1
Resolution	11 Bits
Voltage output	
Output range (rated)	0 to 10 V
Load impedance	<2.5kΩ
Current output	
Output range (rated)	0 to 20ma
Load impedance	<300kΩ

Expansion I/O

Both the PC-50 and the PC-55 can expand the amount of inputs/outputs they can interface to by using 500 Series I/O modules. The following pages describe the various I/O modules available.

The expansion I/O modules connect to the CPU through Bus Units. The bus units are snapped on the mounting rail to the right of the CPU and connected together by a ribbon cable. Each bus unit can hold 2 I/O. Up to 32 I/O modules (6 for the PC-50) can be distributed over up to four mounting tiers using interface modules to interconnect the tiers.

When using 500 Series expansion I/O the CPU and power supply, if used, are first snapped onto the mounting rail. Then, the bus units, into which the I/O modules are plugged later, are snapped onto the rail to the right of the CPU.

The field wiring for the expansion I/O is not connected directly to the I/O module but via a removable screw-type terminal block at the bottom of each bus unit. This allows the I/O modules and bus units to be installed and removed without disturbing the external wiring.

A special keying system helps prevent the insertion of an incorrect module into a bus unit. The rear of each I/O module has a fixed coding "key" depending on the type of module. The bus unit has a rotatable coding element in the form of a "lock". Before a module is inserted, this coding element must be turned to the correct position (corresponding to the marking on the front of the module) to engage the key on the module. Therefore, the bus unit will only take the module type that it is intended for.

I/O Module and Bus Unit Keying

The input/output modules are hooked in the bus units at the top, swung down in the direction of the terminal block and screwed tight to the bus unit. This establishes the connections to both the terminal block and the intenal bus.

Environmental Specifications

Ambient Temperature
- Rail horizontal   0 to 60°C (32 to 140°F)
- Rail vertical   0 to 40°C (32 to 104°F)

Transportation and
storage temperature   −25 to 70°C
                       (−13 to 160°F)

Humidity Rating   15 to 95%, indoors

Mechanical
- Vibrations   IEC 68-2-6
  tested with   10 to 57 Hz
                    (amplitude 0.15 mm)

- Shock   IEC 68-2-27
  tested with   12 shocks
                    (semisinusoidal,
                    15 g, 11 ms)

- Free fall   IEC 68-2-32
  tested with   Height of fall
                    1 m (3 ft.)

Expansion I/O Configurations

PC-50
Max. configuration

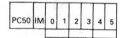

PC-55
Max. configuration
shown with Multi-Tier System
using NLIM-516

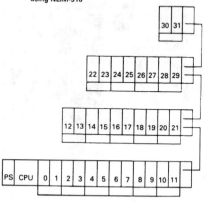

Digital I/O Modules

The 500 Series I O includes a full range of digital inputs and digital outputs with 4, 8 or 32 points per module.

The digital I O modules plug into the bus units. There are no addressing switches on the modules. Addressing is dependent on the location of the bus unit.

Each module is enclosed in a rugged housing which completely protects the electronic components from the environment or rough handling. A wiring diagram is included on the face of each module to assist with field troubleshooting.

Digital Output Modules

Digital Input Modules

Digital Input Modules
The digital input modules convert the external process-level signals to the internal signal level of the programmable logic controller. Choose any standard voltage range from 5 V DC (TTL) to 230 V AC.

Digital Output Modules

The digital output modules convert the internal signals from the CPU to the external signals required to control the process. Modules are available for all standard voltage ranges from 5 V DC (TTL) to 230 V AC. Two relay output modules offer dry contacts for control circuit interlocks.

High Density Digital I/O Module
For those applications where space is at a premium, the NL-572 module provides 16 inputs and 16 outputs in the space of a single 500 Series I O module.

All inputs and outputs are designed for 24 V DC signals. The outputs are divided into two groups of eight. The first group can switch up to 0.1 amp loads. The second group can control up to 0.5 amp loads.

Wiring is to an NLTB-500 terminal block which connects to the front of the module. The NLTB-500 must be purchased separately.

The NL-572 module is addressed as an analog module and can only be plugged into slots 0 to 7. (Slot 0-5 for PC-50 systems.)

NL-539 Relay Output Module with NLTB-500 Terminal Block

NL-572 High Density Module

Technical Specifications

Relay Output Module	NL-529	NL-539
Number of Outputs	4	8
Contact Rating		
• Inductive Load	1.5 A at 250 V AC 0.5 A at 30 V DC	0.5 A at 250 V AC 0.5 A at 30 V DC
• Resistive Load	5 A at 250 V AC 2.5 A at 30 V DC	3 A at 250 V AC 1.5 A at 30 V DC
Rated Switching Operations	1.5×10^6 (AC-11) 0.5×10^6 (DC-11)	1×10^6 (AC-11) 0.5×10^6 (DC-11)
Insulation Voltage	1500 V AC	1500 V AC
External Power Supply • Voltage • Current (without load)	24 V DC 100 mA	24 V DC 70 mA
Heat Dissipation	2 W	1.6 W

Technical Specifications

Digital Input Modules

Digital Input Module	NL-503	NL-505	NL-511	NL-513	NL-515	NL-517	NL-572
Number of Inputs	4	4	8	8	8	8	16
Input Voltage	24-60 V DC	115 V AC	5-24 V DC	24 V DC	115 V AC	230 V AC	24 V DC
Optical Isolation	Yes	Yes	Yes	Yes	Yes	Yes	No
Inputs per Common	4	4	8	8	8	8	8
OFF Voltage Level	−33 to +8 V	0 to 40 V	<25% of L+	0 to 5 V	0 to 40 V	0 to 70 V	0 to 5 V
ON Voltage Level	+13 to 72 V	85 to 135 V	>45% of L+	13 to 33 V	85 to 135 V	195 to 235 V	13 to 30 V
Input Loading	4.5/7 mA	10 mA	—	7 mA	12 mA	16 mA	4.5 mA
Response Time							
• ON to OFF (typ.)	3 msec	10 msec	1 or 10 msec	3 msec	10 msec	5 msec	4 msec
• OFF to ON (typ.)	3 msec	20 msec	(switch selectable)	5 msec	20 msec	15 msec	3 msec
Insulation Voltage	1250 V AC	1500 V AC	500 V AC	500 V AC	1500 V AC	1500 V AC	12 V AC
Heat Dissipation	0.8 W	2.8 W	2.4 W	2 W	2.5 W	2.5 W	4.5 W

Digital Output Modules

Digital Output Module	NL-520	NL-530	NL-531	NL-535	NL-536	NL-572
Number of Outputs	4	8	8	8	8	16
Load Voltage	115/230 V AC	115/230 V AC	5-24 V DC (Sink)	24 V DC (Source)	24 V DC (Source)	24 V DC (Source)
• Operating Range	85-264 V	89-264 V	4.75-30 V	20-30 V	20-30 V	20-30 V
Output Current						
• Per circuit	1 A	0.5 A	0.1 A	0.5 A	1 A	8 x 0.1 A / 8 x 0.5 A
• All circuits	4 A	4 A	0.8 A	4 A	6 A	4 A
Optical Isolation	Yes	Yes	Yes	No	Yes	No
Outputs per Common	4	8	8	8	8	8
Short circuit protection	Fuse 10 A	Fuse 10 A	None	Electronic	Electronic	Electronic
Voltage Drop across output	7 V	7 V	Open-collector	1.2 V	0.8 V	0.8 V
OFF State leakage current	3/5 mA	3/5 mA	0.1 mA	0.5 mA	1.0 mA	0.6 / 1.0 mA
Insulation Voltage	1500 V AC	1500 V AC	500 V AC	—	500 V AC	—
External Power Supply • Voltage • Current (no load)	— —	— —	5-24 V DC 16 mA	24 V DC 15 mA	24 V DC 200 mA	24 V DC 35 / 120 mA
Heat Dissipation	3.5 W	3.5 W	1 W	3.5 W	4 W	4.5 W

Analog Input Modules

The analog input modules convert the analog signals from the process into numeric values for processing in the CPU.

The following can be set by switches on the front of the module:
- 1, 2 or 4 channel operation
- line frequency 50 or 60 Hz
- wire-break alarm signal on/off (not for NL-542)

The analog input modules plug into the bus units. Addressing is dependent on the location of the bus unit and no address displacements occur if modules are interchanged or gaps are left between modules. The analog modules can only be plugged into slots 0 to 7. (Slot 0-5 for PC-50 systems.)

The NL-540 module permits direct connection of up to 4 thermocouples. Junction compensation may be selected for types J, K and L thermocouples. With linearization selected, a measured value of 0°C is transferred to the CPU as 0 units. Each 1°C is equal to 1 unit, greatly simplifying the appliction of temperature inputs.

Analog Input Modules

Technical Specifications

Analog Input Module	NL-540	NL-541	NL-542	NL-543
Input Range	± 50 mV Thermocouple	± 10V	4 to 20 mA	± 500 mV RTD (100Ω Pt)
Number of channels	4	4	4	2
Resolution	12 bits + sign (2048 units nominal value)			
Input Impedance	⩾10MΩ	⩾50kΩ	⩾31.25Ω	⩾10MΩ
External Power Supply				
• Voltage		—	24 V DC	—
• Current	—		80 mA	

Analog Output Modules

The analog output modules convert numeric values from the central controller into the analog signals required for controlling the process.

The analog output modules plug into the bus units. Addressing is dependent on the location of the bus unit and no address displacements occur if modules are interchanged or gaps are left between modules. The analog modules can only be plugged into slots 0 to 7. (Slot 0-5 for PC-50 systems.)

Technical Specifications

Analog Output Module	NL-551	NL-552	NL-553
Output Range	± 10 V	4 to 20 mA	± 20 mA
Number of outputs	2	2	2
Resolution	11 bits + sign (1024 units nominal value)		
Load connection	4-wire	2-wire	2-wire
Load impedance	>3.3 kΩ	<300 Ω	<300 Ω
External Power Supply			
• Voltage	24 VDC	24 VDC	24 VDC
• Current	100 mA	100 mA	130 mA

Analog Output Modules

Analog I/O Resolution

The 500 Series analog I/O modules offer features found only in much larger PLC or process control systems. Wire break detection, 50 or 60 Hz operation, over range, and high resolution are all standard.

For applications requiring precise analog measurement or control, the analog I/O modules should be at least as accurate as the instruments they are connected to. The 500 Series analog input modules utilize 13 bits to represent the measured signal to the CPU. As the table below shows, 1 bit is used for the sign (+ or −) and 12 bits are used for the magnitude of the signal. The analog output modules use 11 bits for the signal magnitude.

NL-541 Analog Input Module

Measured Signal	Analog Value to CPU	
20.000 Volts	>4095	
19.995	4095	Over range
10.0048	2049	
10.000	2048	
5.000	1024	
0.0048	1	
0.000	0	Nominal Range
− 0.0048	− 1	11-bits
− 5.000	− 1024	
− 10.000	− 2048	
− 10.0048	− 2049	
− 19.995	− 4095	Over range
− 20.000	< − 4095	

NL-551 Analog Output Module

Analog Value from CPU	Signal Output	
1280	12.5 Volts	
1025	10.0098	Over range
1024	10.000	
512	5.000	
1	0.0098	
0	0.000	Nominal Range
−1	− 0.0098	10-bits
− 512	− 5.000	
− 1024	− 10.000	
− 1025	− 10.0098	Over range
− 1280	− 12.5	

Slot No	0	1	2	3	4	5	6	7	Channel No
CPU	64 - 65	72	80	88	96	104	112	120	0
	66 - 67								1
	68 - 69								2
	70 - 71	79	87	95	103	111	119	120	3

PC-55 only (columns 6, 7)

Address Assignment for Analog Modules

October 1991

Special Purpose I/O Modules

NL-560 Counter Module

The NL-560 Counter module contains two down counters capable of operation up to 500 Hz. The initial value (1 to 999) is set on the front plate with a thumbwheel switch. The signal level of each input (5 or 24 V) may be set independently by a switch on the front of the module.

Each counter has an enable signal which is set by the program. When set to "1" the preset value is loaded into the counter and the counting of pulses is enabled. The current counter status cannot be read by the CPU.

When the counter reaches 000, its output is set (24 V; can be scanned from the program). The status of the output signal is indicated by a green LED.

The NL-560 appears as a digital I/O module to the CPU and can be plugged into any slot.

NL-561 Counter Module

The NL-561 Counter module contains one counter which can be configured for "counting" or "position decoding" mode by a switch on the front of the module. The signal level of the input pulses can be selected for 5 or 24 V. Operation up to 500 kHz is possible for 5V inputs and up to 25 kHz with 24 V inputs.

The module has an enable input (24 V DC) which must be on to enable counting. The positive edge of the enable signal resets the counter and the outputs.

Two setpoints can be transferred to the module from the user program. A 24 V DC output is turned on when each setpoint is reached. The current count value in the module can be read by the program.

In count mode, the module counts up from 0 to 65,535. For position decoding, the module can count up and down between − 32,767 and + 32,767. The input pulses can be evaluated on the rising edge (single), on the rising and falling edges (double), or on all 4 edges of two pulse streams (quadrature).

Positioning applications can be solved using the NL-561 with an incremental position encoder. For example, the first setpoint can be used to change the drive from rapid traverse to inching speed. The second setpoint can be used to stop the drive at the desired location.

The NL-561 module can only be plugged into slots 0 to 7. (Slot 0-5 for PC-50 systems.)

NL-580 ASCII Communication Module

NL-590 Simulator Module

NL-570 Timer Module

The NL-570 Timer Module contains two electronic timers which can be adjusted over a range of 0.3 to 300 seconds. The delay is preselected for a particular range using a switch on the module front plate and then precisely adjusted using a potentiometer. The CPU starts the timer, which sends a "1" signal to the CPU while the time delay is in progress.

NL-571 Analog Comparator Module

The NL-571 Comparator Module sends a digital signal to the CPU when an external analog signal (0-10 V or 0-20 mA) has exceeded a preset value. The preset value is set with a potentiometer on the front of the module.

An LED on the face of the module lights up when the setpoint is exceeded.

NL-580 ASCII Communication Module

The NL-580 ASCII Communication Module provides bi-directional, serial communication capability for the processor. The module can be operated in print mode or ASCII mode.

Print mode enables output of message texts which are stored on an EPROM/EEPROM memory submodule. Up to 255 messages of 80 characters each can be stored. Each message may include the time and date (from the module's internal clock) and up to 3 variables (transferred from the CPU). Print mode is useful for alarming and event logging to a printer or CRT display.

ASCII mode allows the module to exchange data with other peripheral devices such as terminals, computers, or another NL-580 module. The NL-580 can be used to send and receive both fixed-length and variable-length messages. A memory submodule is not required for ASCII mode.

The NL-580 module contains a real time clock. The clock data maybe read and reset from the program in the CPU. A backup battery (NLB-500) is only required for backup of the clock data during power down.

The NL-580 has a standard 25-pin port for connection of RS-232 or current loop (TTY) devices. The transmission rate can be selected for 110 to 9600 baud.

The NL-580 module can only be plugged into slots 0 to 7.

NL-590 Simulator Modules

The NL-590 Simulator Module is used for simulating digital input signals and for displaying digital outputs. A switch at the rear of the module can be used to select whether the module simulates inputs or outputs.

Like the other I/O modules, the NL-590 plugs into the bus unit. However, it has no mechanical key or wiring connection to the terminal block which allows it to easily replace any digital module. This can be a valuable aid for debugging or troubleshooting.

Description	Catalog Number
Counter Module, 2 channels, up to 500 Hz	NL-560
Counter/Positioning Module, 1 ch., up to 500 kHz	NL-561
Timer Module, 2 timers, 0.3 to 300 seconds	NL-570
Comparator Module, 2 channels, 0-10 V or 0-20 mA	NL-571
ASCII Communications Module	NL-580
NL-580 Operation Manual	NLAM-B501
Simulator Module	NL-590

Power Supplies and Mounting Accessories

NLPS-531 Power Supply

NLR-500 Bus Unit

NLIM-515 Interface Module

Power Supply Modules

The PC-55 has an integral 24 V DC power supply for generating the internal 9 V DC supply voltage. However, if the external mains power source is 115 V or 230 V AC, a separate power supply module must be added to supply the 24 V DC input voltage required by the CPU. An external 24V power supply is not required for the PC-50.

The NLPS-530 power supply should only be used to power the CPU. The NLPS-531 power supply may be used to supply both the CPU and external loads (do not exceed 2A). Both power supply modules plug directly onto the mounting rail. The NLPS-530 is not UL listed or CSA certified.

Short circuit protection of the NLPS-530 power supply is by a 3A fast blow fuse. An electronic means of short circuit protection is provided for the NLPS-531.

Description	Catalog Number
Power Supply for PC-55 only Input 115/230 VAC Output 24 VDC, 1A	NLPS-530
Power Supply for PC-55 and Load Input 115/230 VAC Output 24 VDC, 2A	NLPS-531

Bus Units

The internal I/O bus of the 500 Series programmable controller is assembled by connecting individual bus units. The bus units are snapped side by side onto the mounting rail and interconnected by a flat ribbon cable. Each bus unit can hold two I/O modules. The bus unit contains a removable screw-type terminal block for connecting the external wiring. An I/O module or bus unit may be easily replaced without removing the external wiring.

The mounting rail is a standard DIN rail which may be cut to any length required.

Description	Catalog Number
Bus Unit, mounts 2 I/O modules	NLR-500
Mounting Rail, 19"	NLR-519

Interface Modules

Interface modules are required if the I/O modules are distributed over two, three or four mounting tiers. The interface modules snap directly on the mounting rail and connect to the bus units using the ribbon cable from the bus unit. Interconnection between the interface modules is by a cable.

The NLIM-515, which supports only 1 expansion rail, consists of two interface modules and a 1.6 foot (0.5m) connecting cable.

The NLIM-516 supports 1 to 3 expansion rails. The NLIM-516 modules and cables must be purchased separately. The maximum total cable length is 30 feet (10m).

Description	Catalog Number
Interface for 1 expansion rail, 1.6 ft cable	NLIM-515
Interface Module for 1 to 3 expansion rails	NLIM-516
Cable for NLIM-516, 19" (0.5m)	NLC-505
Cable for NLIM-516, 8.2' (2.5m)	NLC-525

Programming Accessories

NLPL-2605 Programmer

NLPL-2605 Hand Held Programmer

The NLPL-2605 hand held programmer provides an economical means of programming any of the 50 Series controllers. Its
small size makes it a convenient troubleshooting tool. The 2-line LCD display allows
programming and monitoring of programs
in the STL format.

The NLPL-2605 can only be used on-line
connected to the CPU. The programmer
contains internal RAM for up to 1K words of
program statements.

The NLPL-2605 includes a 10 foot (3m)
cable. Operating power is obtained from the
connected CPU.

NLPL-2750 CRT Programmer

The NLPL-2750 is a industrial computer-
based program loader featuring:
- On-line, off-line, documentation software
  for programming in Ladder, Statement,
  and CSF formats
- 80386 processor running at 16 Mhz
- 4 Mbytes of RAM expandable to 16
  Mbytes
- 5.25 in. or 3.5 in. floppy disk drive
- 40 Mbyte hard disk
- 3 spare slots for half-size AT boards, 1 for
  full size board
- EPROM/EEPROM programming facility

Description	Catalog Number
Hand held program loader, 2-line display	NLPL-2605
Carrying Case for NLPL-2605	NLPL-2607
Manual for NLPL-2605	NLAM-B502
Hand held program loader, 8-line display	NLPL-2615
System module for NLPL-2615 (required)	NLPL-2616
Carrying Case for NLPL-2615	NLPL-2617
Accessory power supply for NLPL-2615	NLPL-2618
Manual for NLPL-2615	NLAM-B503

NLPL-2615 Programmer

NLPL-2615 Hand Held Programmer

The NLPL-2615 hand held programmer
offers a flexible, menu driven method of
programming any of the 50 Series controllers. An 8-line LCD display allows programming and monitoring of multiple statements
at one time.

An NLPL-2616 system submodule must be
ordered separately for operation of the
NLPL-2615.

The NLPL-2615 programmer features:
- Eight line LCD display of programs in STL
  format
- Key-operated switch for preventing unin-
  tentional programming
- Internal RAM for 2K statements
- Receptacle for NLEE and NLEP memory
  submodules for transferring programs
  between the internal RAM and the memory submodules
- 10 foot (3m) connecting cable to the programmable controller.

The NLPL-2615 can be operated in two
modes:

- On-line: Connected directly to a CPU, programs can be written and monitored
  directly into the RAM of the controller.

- Off-line: Programs are generated and can
  be stored on memory submodules. This
  mode requires the NLPL-2618 power
  supply.

The NLPL-2615 can be used to transfer programs to the EPROM or EEPROM memory
submodules for the PC-55. The NLPL-2618
power supply is required for EPROM
operation.

Personal Computer

The Westinghouse 3100 Series software
packages convert a Personal Computer into
a versatile program loader. All packages
feature on-line and off-line programming,
full documentation capabilities, print functions, and disk operations.

There are three 3100 Series software packages available for programming 500 Series
controllers:

NLSW-3103 – PC-50 only
NLSW-3104 – 50 Series, 500 Series
NLSW-3105 – 50 Series, 500 Series,
            2000 Series

All packages use identical key strokes and
program files may be easily transferred
between packages.

The software is completely menu driven.
The bottom of the screen indicates the operation of each function key. In addition, over
100 help screens are available to describe
every software function.

Programs may be written and monitored in
either ladder logic or statement list format.
Only a single key stroke is required to
change between formats.

Ladder logic rungs may extend up to 32 elements in series and over 50 elements in
parallel. Rungs can be easily created, edited
and copied.

A properly documented program can eliminate many system maintenance frustrations.
The 3100 Series software makes documentation easy and complete. There are five
methods of annotating a program.

Block Titles provide comments on the function or operation of a block.

Segment Comments describe the function
or operation of a rung.

Statement Comments provides information
about a Statement List instruction or Data
Block register.

Symbols are 30 character labels assigned to
a PLC address.

Symbol Comments are 19 line by 70 character descriptions assigned to a Symbol.

The length of block titles and segment comments are only limited by the amount of
memory available in the computer.

Symbols can be created before or during programming. Once a symbol is created, it can be used in place of the actual PLC address when programming. In ladder format, the symbol appears above the ladder element. In statement list, the symbol appears in place of the operand.

The Data Screen allows monitoring of register data and I/O status. It can also be used for data entry. Any address in the PLC may be displayed. Register data can be displayed in signed and unsigned decimal, hex, ASCII and binary formats.

The printer function supports a variety of options. Print all or portions of a program with output sent directly to a printer or to disk. Printer setup codes can be specified prior to printing.

PROM support is provided through connection of the optional NLEP-2000 PROM programmer. EPROM and EEPROM modules can be read and programmed.

Other features include a complete set of disk operations, search functions and cross reference options.

Computer requirements:
- Personal Computer with 80286 or higher microprocessor
- Minimum 640K of RAM (is expanded or extended memory is supported)
- Standard COM port (serial interface) for communication with the PLC
- LPT port (parallel interface) for connection of the printer and copy protection key
- A hard disk and one floppy disk drive

An RS-232 to current loop converter is required to interface the COM port of the computer to the programmer port of the PLC. The cable and converter are included in the NLCC-3100 communication cable kit.

Each 3100 Series software package includes 3.5″ and 5.25″ disks, copy protection key and manual. A copy protection key is not required for NLSW-3103.

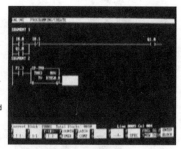

NLSW-3100 Ladder Entry Screen

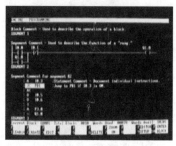

STL and Ladder Display with Documentation

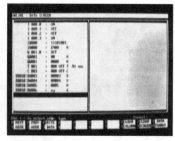

Data Display Screen

Description	Catalog Number
PC-50 Programming Software	NLSW-3103
50/500 Series Programming Software	NLSW-3104
50/500/2000 Series Programming Software	NLSW-3105
Communication Cable Kit	NLCC-3100
PROM Programming Unit	NLEP-2000

Programming Language

The Westinghouse 50 Series, 500 Series and 2000 Series programmable controllers share a common, yet powerful, programming language which goes far beyond the traditional programming methods found in most PLCs. Westinghouse PLCs feature two programming formats, Relay Ladder Logic and Statement List (STL).

The user can choose to program common control functions in familiar Ladder Logic, while also gaining the flexibility and power of a statement based format.

Relay Ladder Logic is familiar to everyone in the electrical controls industry. It is a graphical representation of the PLC program which is very similar to the format used for hardwired electro-mechanical control drawings. Westinghouse PLCs offer the following Ladder functions:

- Contact: N.O., N.C., One-shot
- Coil, Latch, Unlatch
- Timers: Pulse, Extended Pulse, On Delay, Stored On Delay, Off Delay, Reset Timer
- Counters: Count Up, Count Down, Set Counter, Reset Counter
- Comparisons: $=$, $>$, $\geq$, $<$, $\leq$, $><$
- Add, Subtract
- Shift Right, Shift Left (PC-55 only)
- Move, Immediate Input, Immediate Output
- Ones Complement, Twos Complement (PC-55 only)
- AND Word, OR Word, XOR Word
- Jump to block
- Block End, Temporary Block End

Statement List is a text based language which begins with simple logic instructions and extends up to powerful "machine language" type commands. An STL program defines the control logic as a series of steps

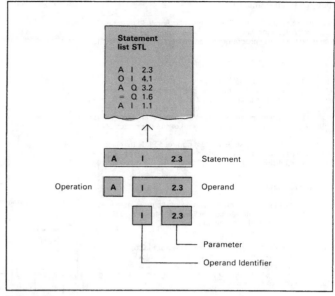

Statement List

or statements. Each statement consists of an Operation (what is to be done) and an Operand (address to act on). Programs written in STL can provide greater flexibility and functionality, with more efficient memory usage, than Ladder programs.

All programs written in Ladder can be easily converted to STL with one key stroke on the program loader. Programs written in STL

can exceed the limits of Ladder notation so may not always be convertible to Ladder.

Since all three Series of controllers use the same programming language and program loaders, programs can be easily transferred to new applications.

Electrical Schematic   **LAD format**   **STL format**

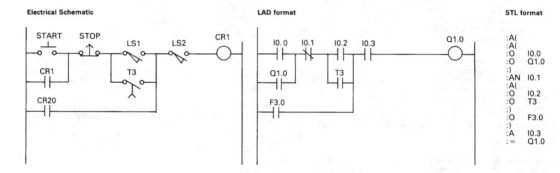

Structured Programming

Westinghouse 50 Series, 500 Series and 2000 Series PLCs offer the control system designer a powerful programming option, the choice between Linear programming and Structured programming.

Linear Programming

Until recently, programmable controllers have only provided Linear programming. In a linear program, the program steps are arranged consecutively, one after another and all instructions in the PLC are processed each scan. As the control sequence becomes more complex, the PLC program becomes longer. The total scan time is also extended because the PLC must process all of the logic each scan.

Linear programming is possible in Westinghouse PLCs by placing all of the program logic in Organization Block 1 (OB1). This block is executed automatically by the PLC each scan.

Structured Programming

For larger applications, the program can be made more manageable if it is broken down into related sections. Westinghouse PLCs include five types of blocks for this purpose.

Organization Blocks (OB) coordinate between the PLC operating system and the application program. Organization block OB1 is executed by the PLC each scan. The first instruction in OB1 is the beginning of the user program. OB1 can be programmed with control logic and calls to other blocks.

Other Organization Blocks provide special functions to make programming easier. For example, the controller only scans the logic in block OB21 one time when the PLC is switched from stop to run mode, before proceeding to the normal scan of OB1. This can be used to initialize variables or alarm an unsafe start condition. Block OB22 is scanned once on controller power up. Block OB34 is only executed each scan that the battery is detected as low.

Program Blocks (PB) normally contain the user application program separated into related sections. Program blocks are created by the user.

Program blocks reside in memory but are only executed when called by the user program. Program blocks may be called unconditionally or conditionally (i.e. depending on logic) and the same block can be called multiple times within the program. Block calls can be stacked up to 16 levels deep.

Function Blocks (FB) are used to create "subroutines" for performing a specific algorithm, data manipulation or calculation. An FB is called in the user program similar to any other Special Function (e.g. timer, counter). However, since the user creates the logic within the FB, it offers the power to create custom special functions.

Function blocks have access to an extended set of STL instructions not available in other blocks. These allow the experienced programmer extensive access to all levels of the controller.

Function blocks can be created with variable names in place of the logic operands. Each time the function block is called, the user can assign actual addresses and data to be used while the function block is executing. Different addresses and data values can be assigned each time the function block is called.

Linear Programming

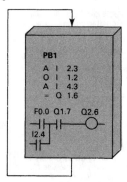

Structured Programming

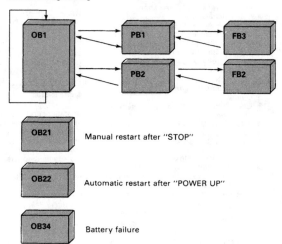

OB21 — Manual restart after "STOP"

OB22 — Automatic restart after "POWER UP"

OB34 — Battery failure

A function block must be given a name of up to eight characters. When the FB is called, the assigned name appears with the function. This makes it easy to identify what function the FB block performs.

One advantage of function blocks is that the programmer has the ability to create special functions that do exactly what is needed, in a language that is best suited to high level data manipulation. At the same time, to the maintenance person looking at the Ladder logic call of a function block, it appears the same as other familiar special functions such as timers and counters.

Data Blocks (DB) are created by the user as needed for data storage. Each DB can consist of between 2 and 256 16-bit data words (DW). Each data word may contain integer, hex, ASCII or binary data.

Operator Interface Panels

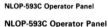

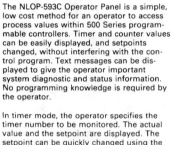

NLOP-593C Operator Panel

Timer Display   Message Display

NLOP-593C Operator Panel

The NLOP-593C Operator Panel is a simple, low cost method for an operator to access process values within 500 Series programmable controllers. Timer and counter values can be easily displayed, and setpoints changed, without interfering with the control program. Text messages can be displayed to give the operator important system diagnostic and status information. No programming knowledge is required by the operator.

In timer mode, the operator specifies the timer number to be monitored. The actual value and the setpoint are displayed. The setpoint can be quickly changed using the numeric keypad. The arrow keys can be used to scroll through the timers.

Counter mode operates similarily to timer mode. In both modes, the system programmer can prevent the operator from changing setpoints.

Message mode permits up to 128 messages of 16 characters to be displayed on the operator panel. The message texts are stored in the CPU. If several messages are pending simultaneously, the message with the highest priority is displayed first. The other messages can then be called (in decreasing priority) using the scroll key.

Designed for industrial environments, the NLOP-593C consists of a watertight case with the keys and display on a watertight front. The Operator Panel is small enough to be operated as a hand held unit or flush mounted in a control panel (using the optional mounting frame).

The NLOP-593C connects to the CPU programming port via a 10 foot cable (included). Operating power is obtained from the connected CPU. The mounting frame must be ordered separately.

Description	Catalog Number
Operator Panel, Timer/ Counter/Data access	NLOP-593C
Mounting frame	NLOP-593F
NLOP-593 Operation Manual	NLAM-B505

Dimensions

NLOP-593B Operator Panel

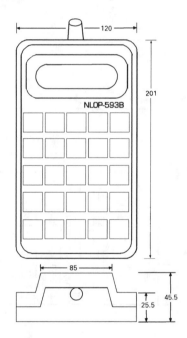

Conversion table mm to inches

mm	25.4	45.5	85	120	201
in.	1	1.8	3.34	4.72	7.91

PC-50 Dimensions

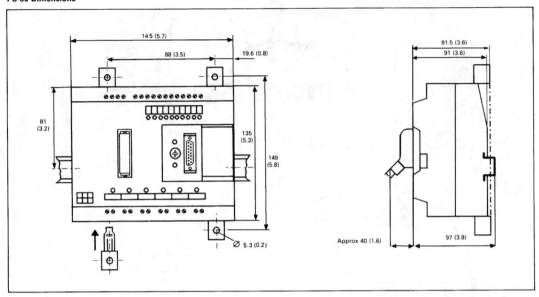

PC-55 Dimensions

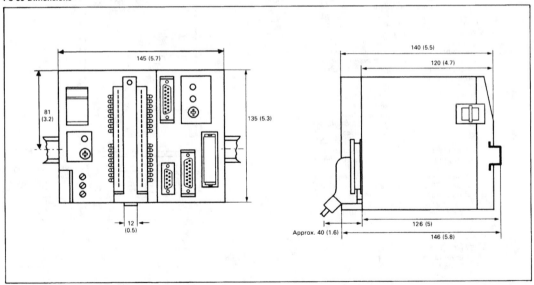

Product Support

Local Distribution
In over 70 locations around the United States and Canada, full service Westinghouse programmable controller distributors stock a complete line of 2000 Series components. Westinghouse full service distributors can also provide training in the operation and maintenance of PC-2000 systems.

Application Support
Located in key Westinghouse sales offices around the United States and Canada, programmable controller specialists provide technical assistance in the specification and use of Westinghouse programmable controller products.

Factory Assistance
Westinghouse also provides factory application and technical assistance to its customers. Available by telephone, Westinghouse personnel quickly respond to customer needs, including troubleshooting, analyzing system operation, and coordinating component repair or replacement.

Factory assistance may be obtained by telephoning 800-542-7883 or 412-937-6790.

Training
Westinghouse provides comprehensive training on all aspects of Westinghouse programmable controllers from its Pittsburgh Training Center. Courses include system configuration and design, programming, troubleshooting and maintenance. The Training Center also offers on-site training for all its courses.

Westinghouse also provides programmable controller training in conjunction with its Full Service Distributors.

Further Information

Medium systems

Reprinted with the permission of Eagle Signal Controls, a division of Mark IV Industries, Inc.

The Eagle Series Programmable Controllers Precision Machines That are Easy to Use

The **Eagle** Series of Programmable Controllers includes the **Eagle-1**, **Eagle-2**, and the **Eagle-3**. All the controllers provide standard, built-in features that offer unsurpassed power, flexibility, and ease-of-use.

Powerful fill-in-the blanks programming

The **Eagle** Series Programmable Controllers use simple-yet-powerful programming software called **ESRL** (Enhanced Symbolic Relay Language), combined with your personal computer, to develop ladder logic diagrams. Extensive use of fill-in-the blanks function blocks allow the user easy access to the full power of the controller, including PID and network configuration. The on-line HELP function, available at a single keystroke, greatly reduces the time needed to program the **Eagle** Series Controller.

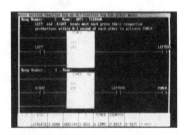

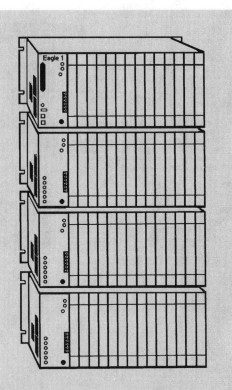

- Enhanced **SRL** (Symbolic Relay Language) Programming/Monitoring
- Up to 2048 I/O, Local and Remote
- Local and Remote Operator Interfaces
- Up to 64 PID Loops
- Recipe Storage
- 0.75 mSec/K Scan Time
- Supplemental Software Routines for Links to BASIC, C, and Other Languages for User-Defined Data Acquisition

Enhanced SRL Programming — Software With the User in Mind

Eagle Signal's new Enhanced **SRL** (Symbolic Relay Language), makes programming even complex PID loops as simple as filling in the blanks of the screen shown below:

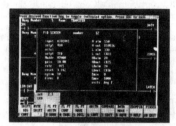

Most of the data fields shown can be changed "on line" allowing the systems engineer or programmer to tune his machine or process without reprogramming or halting the system. Each PID loop can be monitored in real time, with changing values shown on the screen. Complex control strategies are possible using analog data and math calculations in the logic of the process.

All **Eagle** Series Controllers have built-in report generation capability, allowing formatting and printing from within the controller's memory. Once a report has been created with **EZ TEXT** (see the example below), a printer or dumb terminal can be connected, freeing the user's computer.

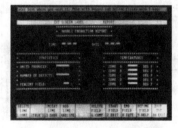

Some key features of Eagle's Enhanced **SRL** include:

- Fill-in-the-Blanks Function Blocks for Timers, Counters, PID Loops, and Communication Networks
- On-Line Tuning of PID Loops (Eagle 2 and 3)
- **EZ TEXT** Report Generation
- On-Line **HELP** Screens
- Built-In **COMMENTS** Enhance Documentation Without Third Party Software
- Recipe Storage

The **Eagle 1** Programmable Controller offers a midsize system of unsurpassed flexibility and ease-of-use.

Using field-proven technology, with a highly robust power supply and noise-immune components, **Eagle** Series Controllers are at home in any environment. The **Eagle 1** includes the following features:

- 896 Local Digital I/O
- 125 Timers, Including .01 Second High-Resolution
- 100 Counters
- 1000 Control Relays
- 500 Retentive Control Relays
- 6 High-Speed Counters
- 1000 16 Bit Fixed-Point Data Registers
- **Recipe Storage in Plug-A-PROM**
- **Immediate I/O Update** (Not Scan-Dependent)
- **0.75 mSec Scan Time** (Per 1K Program)
- Complete line of AC and DC I/O, Some with 2A Rating
- 2 RS-232C Serial Interfaces
- Up to 38K of User Program/Data Memory, Either Battery-Backed RAM or UVPROM
- **On-Line Diagnostics of CPU Operation**
- Individually Fused Outputs
- Built-In Networking, Communications and Distributed Control Capability
- Remote (2000 ft.) **DADM** (Data Access and Display Module) and Time-of-Day Clock

In addition to all the capabilities of the **Eagle 1**, the **Eagle 2** has the following features:

- 64 Analog I/O, 12 Bit Resolution
- 32 Pre-Defined PID Loops, Each With Dedicated, Configurable Setpoint and Deviation Alarms, and On-Line Monitoring
- 32 Bit Floating Point Math With Transcendental and Exponential Functions
- Automatic Linearization of Thermocouple Inputs
- Bumpless Auto-Manual Transfer
- 100 32 Bit Floating-Point Data Registers

The Eagle 3. Where Precision Control Needs Power You Can't Find Anywhere Else, Eagle 3 is the Answer.

The **Eagle 3** Programmable Controller gives the user precision control solutions to any complex problem, in any situation. Whether your need is water/waste; high-speed machine; petroleum or petrochemical product refining, processing, distribution or trasmission; hazardous material management; simple or complex networks, **Eagle 3** offers the power, flexibility and ease-of-use you require.

All **Eagle** Series Programmable Controllers share a common programming language. This means that you start with only what you need, and as you grow, the **Eagle** grows with you. Any program written for an **Eagle 1** or **2** will run unchanged in an **Eagle 3**.

The **Eagle 3** offers all the capabilities of the **Eagle 2**. It goes far beyond the traditional mid-size controller, however, with features that are unequalled at comparable cost. These features include:

• 2048 Total I/O, Local and Remote
• 1144 Total Analog I/O, Local and Remote
• Up to 512 Total Remote Analog I/O
• 64 PID Loops with Alarms
• Auto-Manual PID with Bumpless Transfer
• Fast PID Scan Times Under Program Control
• 12 High-Speed Counter Inputs
• 8 RS-232C Serial Interfaces
• 187 KBaud Remote I/O

The **Eagle 3** can be configured as the central "host" in a distributed control system, eliminating the need for expensive, dedicated computer-based systems. With Eagle Signal's **ECOM** network, stand-alone controllers provide overall system integrity, allowing critical installations to stay "on line" if communication with the central controller fails.

The **Eagle 3** Programmable Controller, like the EPTAK 7000, supports up to four chassis, and the Eagle Remote I/O system. New for both controllers is remote chassis I/O, allowing more remote I/O control in less space. You can now install up to 224 I/O in a 19″ rack-mounted chassis, up to 10,000 feet from the CPU. This I/O can be a combination of digital and analog, with the remote analog having full PID capability. Up to eight remote I/O drivers can be installed in an **Eagle 3** or EPTAK 7000, each driver capable of controlling up to 256 total remote I/O. As specified, the combined total of all local and remote I/O, analog and digital, is 2048.

The **Eagle 3** allows up to eight RS-232C serial interfaces per system. This provides you with the ability to connect printers and dumb terminals, and to monitor and control peripherals such as:

● Intelligent Servo Drives
● Intelligent Stepper Motors
● Intelligent Variable Speed Drives
● Intelligent Weigh Scales
● Process Analyzers

A possible **Eagle 3** configuration:

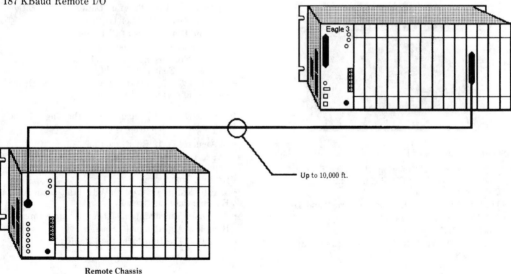

Up to 10,000 ft.

Remote Chassis

Local I/O

**Located in Processor Chassis or in an
Expansion Chassis**

Local I/O — Discrete

Discrete I/O modules contain up to 16 optically
isolated circuits per module, up to 224 I/O per
19″ chassis.

Inputs
5, 12, 24, 48 VDC
24, 48 VAC
120 VAC

Outputs
2 Through 55 VDC
24, 48 VAC
120 VAC (2A)

Local I/O — Analog

Both input and output modules have 12 bit resolution.
Inputs have 8 circuits per module; outputs have 4
circuits per module.

Inputs
Direct Thermocouple (J, K, R, S, T, E)
Direct RTD (100 Ohm Platinum, 3W)
Voltage (1-5, 0-5, 0-10, ± 10)
Current (4-20 mA, 10-50 mA)
0-500 mV

Outputs
Voltage (0-5, 0-10, ± 5, ± 10)
Current (4-20 mA, 10-50 mA)

Local I/O — Special Purpose

BCD/Binary/TTL
High-Speed Counter (15 KHz), 6 circuits
RS-232C Interface

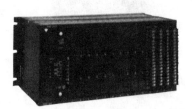

Remote Chassis

Remote Track

Remote I/O

The Eagle 3 Programmable Controller supports the same remote I/O System as the EPTAK 7000 Industrial Control Computer. Either dense chassis I/O with the CP7035, or 2-point track I/O with the CP2035 can be used with the Eagle 3.

CP7035 Remote Chassis I/O

The CP7035 Remote I/O Chassis includes 14 slots, for a total of up to 224 I/O in 19″ rack or panel mounting. Up to 64 analog I/O can be installed, providing significant savings when multiple temperature points must be located remotely from the central control room. Only a single shielded, twisted pair, up to 10,000 feet, transmits all I/O data for monitoring and control.

CP2035 Remote 2-Point I/O

Remote I/O flexibility is greatly increased with the 2-point I/O system, originally developed for Eagle Signal's EPTAK 200 Series of small controllers. Using the same remote I/O driver module as the CP7035 Remote Chassis, the CP2035 Receiver and the CP2050 Track provide up to 32 I/O, in 2-circuit modules. Eagle's remote I/O network operates at 187 KBaud, allowing fast update of both analog and digital I/O status.

2-Point Discrete I/O

Inputs
24, 48, 120, 240 VAC
10-55 VDC
5 VDC TTL

Outputs
120, 240 VAC (N.O. or N.C.)
10-55 VDC
0-250 VDC Universal
120 VAC 5A SPDT Relay
120 VAC 10A DPDT Relay

Special Purpose
I/O Simulator
Circuit Test
Watchdog Timer

2-Point Analog I/O

Inputs
Direct Thermocouple (J, K, R)
Direct RTD (100 Ohm Platinum, 3W)
Voltage (0-10, 0-5, 1-5)
Current (4-20 mA, 10-50 mA)

Outputs
Voltage (0-10, 1-5)
Current (4-20 mA, 10-50 mA)

The **Eagle** Series Programmable Controllers provide the user with a variety of peripheral devices and supplementary software that allow configuration for any application.

Operator Interface Terminals

The **CP9200** (monochrome) and **CP9240** (color) Operator Interface Terminals provide a fully interactive interface for machine or process view. These rugged, industrially-hardened NEMA-4/12 terminals feature:

- Built-in **ECOM** Communications
- Interactive Graphics for Machine/Process View
- Networking to Communicate with Multiple Processors
- Easy, Built-in Programming Language
- Printer Port for Alarm or Data Logging
- Real Time, On Line Trending

Message Center

For use where good readability and visibility are essential, Eagle's **CP9110** Message Center can replace conventional pilot lights, meters and switch panel instrumentation. The **CP9110** can be used for operator prompting, production or process status reporting, alarm messages, or to display system status or condition. Messages can be programmed into its memory and displayed and/or printed. Key features include:

- 2 Line, 20 Character Vacuum Fluorescent Display
- 400 Programmable Messages for Alarms, Data Display, Instructions or Information, Operator Prompting
- Serial or Parallel Input
- Printer Interface
- Easy Programming, Menu-Driven Editor

DADM
Data Access & Display Module

These compact, panel-mounted modules allow timer, counter and analog setpoint changes, and monitoring of PID, timer and counter "actual" values. Eagle's **DADMs** (Data Access & Display Modules) feature:

- Remote Location (Up to 2000′ from the Controller)
- Up to 9 **DADMs** can be used with a Single Controller
- Keylock Security Prevents Unauthorized Operator Access

Personal Computer

Your IBM, Compaq or 100% compatible computer provides an excellent "window" to your machine or process, when combined with Eagle's **ESRL** or other supplementary software, or with a variety of interactive graphics packages. It can also be used as a "dumb" terminal. Currently available graphics and data logging packages include:

- QBECOM® from Eagle Signal
- FactoryLink® from US Data
- ONSPEC® from Heuristics
- CIMPAC® from Action Instruments
- The FIX® from Intellution

Component Descriptions

Description	Catalog Number
Eagle 1 4-Slot Controller	CP7404
Eagle 1 7-Slot Controller	CP7407
Eagle 1 14-Slot Controller	CP7414
Eagle 2 4-Slot Controller	CP7504
Eagle 2 7-Slot Controller	CP7507
Eagle 2 14-Slot Controller	CP7514
Eagle 3 4-Slot Controller	CP7604
Eagle 3 7-Slot Controller	CP7607
Eagle 3 14-Slot Controller	CP7614
Expander Chassis, 7-Slot	CP7907
Expander Chassis, 14-Slot	CP7914
RAM Memory, 16K	CP7750
RAM Memory, 32K	CP7751
RAM Memory, 48K	CP7752
PROM Memory, 48K	CP7760
Redundant Power Supply, 120 VAC	CP7651
AC or DC Input (16)	CP729
AC or DC Output, 2A Indiv. Fused (16)	CP739
12/24/48 VDC Input (16)	CP722
24 VAC Input (16)	CP723
24/48 VDC Output (16)	CP732
24 VAC Output (16)	CP733
BCD/Binary I/O (16/16) (Opt. Isol.) (Specify Voltage)	CP728
A/D Converter	CP744
Analog Calibration Card	CP747
Thermocouple Input (8) (Specify J, K, R, S, T, E)*	CP741
RTD Input (8), 100 Ohm Plat., 3W*	CP743
Analog Input (8), 1-5V, 4-20mA*	CP754
Analog Output (4), 4-20mA & 0-5V*	CP756
Analog Output (4), 0-10V*	CP757
High-Speed Counter (6), 15KHz	CP724
Data Entry/Display (6 Slots)	CP735
Pushbutton Module (4)	CP725
RS-232C Serial Interface	CP717
Error Indicator Module	CP737
Watchdog Timer Module	CP731
Data Bus Extender/Indicator Card	CP704

* 12 Bit Resolution

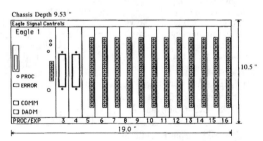

Chassis Depth 9.53 "

4-Slot Controller Width = 9 "   7-Slot Controller Width = 12 "

Remote I/O System — Eagle 3 Only

Description	Catalog Number
Remote I/O Driver Module	CP716
14-Slot Remote I/O Chassis	CP7035
Remote I/O Receiver, Track Mounting	CP2035
Remote I/O Track (32)	CP2050
Remote I/O Track (16)	CP2051
110 VAC/DC Input[1]	NL110
240 VAC Input	NL111
24 VAC Input	NL112
10-55 VDC Input	NL130
TTL Input, 5 VDC	NL131
120 VAC Output, N.O.	NL120
240 VAC Output, N.O.	NL121
10-55 VDC Output	NL140
Relay Output, DPDT, Single Coil, 10A	NL160
Relay Ouptut, SPDT, Sol. State, 100mA	NL161
Relay Output, SPDT, Mech., 5A	NL162
Analog Input, 1-5V	NL230
Analog Input, 4-20mA	NL231
Analog Input, 10-50mA	NL232
Analog Input, 0-5V	NL233
Analog Input, 0-10V	NL234
Analog Input, RTD	NL237
J Thermocouple Input[2]	NL280
K Thermocouple Input	NL281
Analog Output, 0-10V	NL240
Analog Output, 4-20mA	NL241
I/O Simulator Module	NL150
Circuit Test Module	NL151
Watchdog Timer Module	NL170

[1] All "NL" Modules except NL160 contain two independent isolated circuits.

[2] Other thermocouple inputs are available on special order. Consult your authorized Eagle Signal Deale for pricing and availability.

CONFIGURING YOUR SYSTEM

Step 1: Determine the type and quantity of input, output, and special purpose modules the application requires.

Step 2: Choose the controller type and size based on the required modules.

Step 3: Select peripheral devices such as DADMs, Operator Interface Terminal, Message Display Center, etc.

Specifications

Networking Communications Distributed Control	Any of the Eagle Series may be configured as distributed, "stand-alone" controllers from a central EPTAK 7000 system or host computer. See Eagle Signal Bulletin 5005-780 for details of network implementation.		
DESCRIPTION	EAGLE 1	EAGLE 2	EAGLE 3
Operating Temperature	0-60°C	0-60°C	0-60°C
Humidity	95% Non-Condensing	95% Non-Condensing	95% Non-Condensing
Voltage	120/240 VAC, 24 VDC	120/240 VAC, 24 VDC	120/240 VAC, 24 VDC
Scan Time: mS/K Program	0.75 mS	0.75 mS	0.75 mS
Memory User Program/Data (Max.) Type	38K BBRAM, UVPROM	38K BBRAM, UVPROM	38K BBRAM, UVPROM
Timers 0.1 Second 0.01 Second	100 25	100 25	100 25
Counters	100	100	100
Data Registers Fixed-Point, 16 Bit Floating-Point, 32 Bit	1000	1000 100	2000 100
Control Relays (Retentive)	500	500	500
Control Relays (Non-Retentive)	1000	1000	1000
Byte Shift Registers, 255 Step	8	8	8
High Speed Counters	6	6	12
PID Algorithm	-0-	32	64
Digital I/O Maximum	896	896	2048
Analog I/O Maximum	-0-	64	1144
Digital Inputs-Outputs Circuits/Module 24, 48, 120, 240 VAC 24, 48 VDC BCD/Binary/TTL	16 16 16, 32	16 16 16, 32	2, 16 2, 16 2, 16, 32
Analog Inputs/Module (12 Bit) Voltage Current Thermocouple (J, K, R, S, T, E) RTD (100 Plat., 3W)		8 8 8 8	2, 8 2, 8 2, 8 2, 8
Analog Outputs/Module (12 Bit) Voltage Current		4 4	2, 4 2, 4
Remote I/O System Digital, Analog, PID Max. Distance from CPU Communication Speed I/O Points Per Housing/Rack			Yes 10,000 Feet 187 KBaud 16, 32, 224

A Full Line of Control Products Backed by a Worldwide Network of Industrial Control Specialists

From the micro-size EPTAK 100 through the advanced EPTAK 7000, Eagle Signal Controls offers a complete family of powerful control systems. For more information, contact the Authorized Eagle Signal Controls Dealer nearest you, or call **Eagle Signal Controls** (512) 837-8300.

Large systems

Reprinted with permission of UTICOR Technology, Inc.

DIRECTOR 6001 PROGRAMMABLE CONTROLLER

FEATURES

- 68000 Microprocessor based Solver Board
- 68000 Microprocessor based Scanner Board
- 64K, 128K, or 256K User CMOS RAM
- Two I/O Channels (32 Tracks per Channel, 2048 I/O per Channel, 4096 I/O TOTAL)
- 1024 TIMERS (.01 to 10 Second Time Base)
- 21000 DATA Registers (16 bit) less Table/Block definition area
- COUNTERS-Any DATA Register location
- 256 DATA Inputs and 256 DATA Outputs (16 bit)

- 256 Analog Inputs and 256 Analog Outputs
- Self Diagnostics on Power-Up
- Down-Loadable EXECUTIVE Program (optional)
- Real-Time Clock
- RS-232/RS-422/RS-485 Serial Communications
- High Speed I/O Interface (100K or 200K BAUD)
- PROGRAM LOADER software will run on any fully compatible IBM PC
- OFF-LINE/ON-LINE EDITING
- Documentation for PROGRAM, PAGE, & ELEMENT

The DIRECTOR 6001 Programmable Logic Controller (PLC) provides a large user program (up to 256K bytes), executes the user programs at a very high rate (2.45 mSec for every 1000 relay elements), and provides a large I/O structure (4096 I/O circuits).

The DIRECTOR 6001 PLC system consists of the processor, high speed I/O track interfaces and program loader software.

The processor boards are IBM AT form factor with an AT buss edge connector and interface circuitry to allow them to be plugged into a PC, XT or AT buss.

The DIRECTOR 6001 uses a high speed I/O track interface. All of UTICOR's high density I/O modules and high density tracks are compatible and both discrete and data modules may be intermixed on the same track.

IBM is a registered trademark of International Business Machines Corp.

Memory Size	:	64K, 128K, 256K Bytes
Dimensions	:	18" X 8 1/2" X 3"
Front Panel Indicators	: :	Green "LOGIC POWER" LED Green "BATT. STATUS" LED Green "RUN STATUS" LED Green "COMM. ACTIVE" LED Red "CPU FAULT" LED Red "MEM. FAULT" LED Red "I/O FAULT" LED Red "I/O FORCED" LED
Front Panel Controls	:	3 position key-switch controls modes of RUN, REMOTE, and PROGRAM
Service Power	:	115 VAC (102-132) 47-63 HZ 230 VAC (194-250) 47-63 HZ
Operating Temp	:	0 to +60 Degrees C ambient

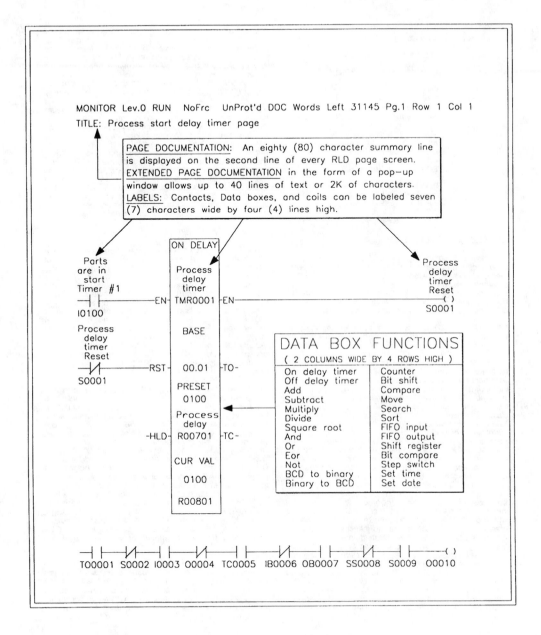

MONITOR Lev.0 RUN   NoFrc   UnProt'd DOC Words Left 31145 Pg.1 Row 1 Col 1

TITLE: Process start delay timer page

PAGE DOCUMENTATION: An eighty (80) character summary line
is displayed on the second line of every RLD page screen.
EXTENDED PAGE DOCUMENTATION in the form of a pop-up
window allows up to 40 lines of text or 2K of characters.
LABELS: Contacts, Data boxes, and coils can be labeled seven
(7) characters wide by four (4) lines high.

Parts
are in
start
Timer #1
—| |—   —EN—

I0100

Process
delay
timer
Reset
—|/|—   —RST—

S0001

ON DELAY

Process
delay
timer
TMR0001 —EN—

BASE

00.01   —TO—

PRESET
0100

Process
delay
R00701   —TC—

CUR VAL

0100

R00801

Process
delay
timer
Reset
—( )—

S0001

—HLD—

DATA BOX FUNCTIONS
(2 COLUMNS WIDE BY 4 ROWS HIGH)

On delay timer	Counter
Off delay timer	Bit shift
Add	Compare
Subtract	Move
Multiply	Search
Divide	Sort
Square root	FIFO input
And	FIFO output
Or	Shift register
Eor	Bit compare
Not	Step switch
BCD to binary	Set time
Binary to BCD	Set date

—| |— —|/|— —| |— —|/|— —| |— —|/|— —| |— —|/|— —| |———( )—
T00001 S0002 I0003 00004 TC0005 IB0006 OB0007 SS0008 S0009 00010

Storage Temp:		-40 to +95 degrees C ambient
Humidity	:	10 to 95% RH Non-Condensing
Shock / Vibration	: :	Vibration Excursion: Double Amplitude .060" Peak to Peak Frequency: 10-55 HZ Linear Ramp Duration: 30 Minutes each axis Shock: 1/2 Sinewave @ 11 mSec duration Peak Acceleration: 40 G each axis
Electrical Noise Tolerance	: : :	NEMA ICS 2-230 Showering Arc Test ANSI C37.90A-1974 (SWC) Surge Withstanding Capability Test SAMA PMC 33.1 EMI/RFI Test Radio Frequency Test: 5 WATTS @ 70-75 Mhz 5 WATTS @ 140-145 Mhz

The CENTRAL PROCESSOR

The **DIRECTOR 6001** may be used in any automation system that requires a large I/O structure, large user memory, and rapid execution of the user program. The physical dimensions (18" X 8 1/2" X 3") are identical to those of the DIRECTOR 4001 to allow existing systems to be upgraded without rework. It's inside where the differences exist, both physically and functionally.

FUNCTION • The Director 6001 has a sixteen (16) bit 68000 microprocessor on each of its boards. The Solver board contains 64K, 128K or 256K bytes of battery-backed user memory and the three (3) seperate nine pin "D"-type female serial communication ports. All three ports can communicate from 300 BAUD to 19.2K BAUD. The I/O Scanner board contains the system memory, the real-time clock and a single in-line connector providing the two (2) serial ports for the RS-485 serial communication links to the high speed I/O tracks at 100K or 200K BAUD. I/O communication timers guard against a communication failure. In the event of a failure, the I/O track interface will either disable or hold the I/O. The processor will then set a scratchpad address for use in the user program. Each serial link supports up to 32 I/O tracks. Both boards utilize watchdog timers to insure hardware and software failures are detected. The Solver board executes the user program while the Scanner board scans the field I/O. As long as the I/O scan time is shorter than the relay ladder execution time, it will not impact the total scan time of the processor. Operating together, they exchange information packets, with the Solver updating its program with the I/O information and then telling the Scanner to update the system I/O. In this manner the scan rate of the DIRECTOR 6001 and the operating time of the system is significantly reduced. Program execution speed is 2.45 mSec typical for every 1000 relay elements. The I/O snapshot time is dependent not only on the number of I/O, but on the number of I/O per serial link as well as the mix between discrete and data modules. The snapshot time for 304 discrete I/O (38 modules) and 38 data modules with the modules divided evenly between the two serial links running at 200 KBAUD is 6 mSec.

DIAGNOSTICS • Several tests are performed when power is first applied or when the processor is restarted. The results of the tests are stored in a register. If any of the tests fail, the processor will not enter the RUN mode.
-RAM memory is tested by a read/write test pattern to all system RAM except the retentive area.
-The USER PROGRAM is tested with a cyclic redundancy check (CRC). A checksum is computed on the memory and compared with a stored value.
-PROM is also tested by a CRC. A computed checksum on this memory is compared to a stored value.
-The SYSTEM HARDWARE is tested by functional tests on the processor boards for proper operation.
-HARDWARE/SOFTWARE CONFIGURATION is provided for every board in the processor as a means of determining the board's hardware revision, software revision, and memory configuration. The memory configuration includes the amount of each type of memory and the type of chips installed. This configuration data is set-up at the factory and protected against unauthorized alterations. This data is stored in the system memory and can be accessed via the program loader or serial interfaces.

The SYSTEM

The DIRECTOR 6001's two I/O ports communicate with **HIGH SPEED I/O INTERFACE** units. The High Speed I/O Interface is comprised of an intelligent interface board, a power supply for powering the interface board, and I/O modules. The interface board is 6809 microprocessor based and provides I/O scan data, module status, and fault information to the processor over a serial link. The serial link is RS-485 and communicates at either 100K or 200K BAUD for cable distances of 4000 feet or 2000 feet respectively. Discrete I/O and Data I/O can now be intermixed on any track in the system. The High Speed I/O Interface mounts directly on the I/O track and when upgrading from a DIRECTOR 4001 will be a drop in replacement for existing UTICOR high density I/O interface units.

An **Industrial IBM** or fully **IBM compatible personal computer (PC)** serves as the program loader with the support software from UTICOR designed just for the D6001. With an IBM PC as the program loader, serial communication is RS-232 with the baud rate varying from 300 Baud to 19.2K Baud. If the PC has RS-422 capabilities, two hundred fifty five (255) DIRECTOR 6001s may be accessed from a single IBM PC. The D6001 Program Loader Software uses conventional Relay Ladder Logic. Data Boxes are placed in with the relay pages. The program page format contains ten (10) rows and each row consists of ten columns (data boxes occupy 4 rows by 2 columns). Each element's address is prefixed with a code letter(s) that indicates what type of element it is, (i.e. O=output, I=input, S=scratchpad). Each page, element, and coil can be labeled for easy identification. In addition, the entire program can be identified by creating a **Header Document** of 255 lines or 10000 characters.

Programming is protected by a combination of 5 **Protection Levels** (0-4) and a **Password** that is set from the highest level (Protection Level #4). Each level, with the exception of Level #0, allows varying degrees of access to the DIRECTOR 6001's user program. Level #0 only allows monitoring of the user program but does not allow any changes to be made. Programs that are tampered with

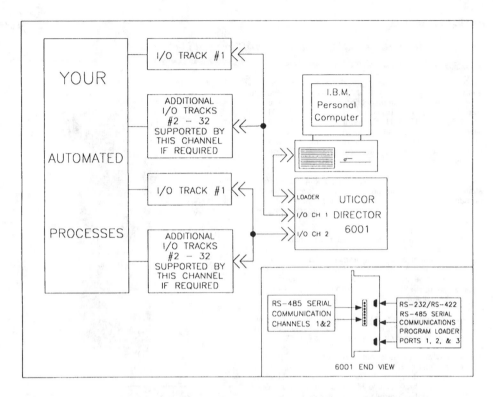

YOUR

AUTOMATED

PROCESSES

I/O TRACK #1

ADDITIONAL
I/O TRACKS
#2 – 32
SUPPORTED BY
THIS CHANNEL
IF REQUIRED

I/O TRACK #1

ADDITIONAL
I/O TRACKS
#2 – 32
SUPPORTED BY
THIS CHANNEL
IF REQUIRED

I.B.M.
Personal
Computer

LOADER   UTICOR
I/O CH 1   DIRECTOR
I/O CH 2   6001

RS–485 SERIAL
COMMUNICATION
CHANNELS 1&2

RS–232/RS–422
RS–485 SERIAL
COMMUNICATIONS
PROGRAM LOADER
PORTS 1, 2, & 3

6001 END VIEW

can readily be identified by a comparison of **Time Stamps**, where the time stamps in the controller are compared to the time stamps of the same program filed on disk. Additionally, the D6001 keeps a log of Protection Level changes.

ON-LINE monitoring has been enhanced with the addition of **HISTOGRAMS**. Three Histograms of 8 discrete elements each are provided to allow the tracing of elements' states for 255 scans after the trigger condition is met. A visual graph is stored and displayed and can be used to assist in troubleshooting the system for timing problems or examining that particular events occur within a given time frame.

The D6001 **FORCE ADDRESS** function is unique, in that, no address can be forced until the **ENABLE FORCE** function is invoked. Then, when all the addresses that need to be forced, are, the D6001 can release them all at once by either the use of the **SYSTEM FORCE RELEASE**, or the **DISABLE FORCE** function. The advantage of the Disable Force is that the forced state of each element is retained

and all will return to their forced state when Enable Force is selected again. Of course, each element can still be individually forced on/off.

The DIRECTOR 6001 opens the door to a new approach on automation with features like the **DOWN-LOADABLE EXECUTIVE OPTION** which allows executive programs to be up-graded, in-circuit, via the program loader or serial interfaces when revisions are made.

The DIRECTOR 6001 is capable of providing an interface between many types of programmable logic controllers and, in the future, will allow integration of more controllers for the next generation of automated process systems.

ORDER INFORMATION

To order the DIRECTOR 6001, contact your authorized local distributor or call UTICOR Technology, Inc. at (319) 359-7501.

Peripherals

Reprinted with permission of Square D Company

SY/NET Transfer Network Interface Modules (TNIM) enable two devices on the SY/NET Local Area Network to share a common communication route. This capability is beneficial in applications using the SY/MAX® Local/Remote (LTI/RTI) Transfer Interface System in which a pair of processors (primary and backup) control a common I/O system, but where communication is normally required only with the controlling (primary) processor.

The TNIM allows the primary and backup processor to share a communication route, but only allows the current primary processor to use the shared or "primary" route. This provides the following communication system advantages:

1) Enables fewer communication instructions for any device on the SY/NET network that communicates to the transfer system controllers. Because the device does not need to determine which is the primary and backup controller, the need for different communication rungs is eliminated.

2) Communication instructions for the primary and backup controller can be identical, thereby simplifying their control programs.

TNIMs may also be used with a pair of devices in a non-LTI/RTI configuration to enable automatic transfer of their communication function. These devices include the SY/MAX Class 8020 Deluxe Model 300 Processor, Class 8030 D-LOG Data Controllers, computer or similar device with SY/MAX protocol.

SY/MAX Transfer Interface System

In a basic SY/MAX Transfer Interface System, a pair of SY/MAX Model 400, 500, 600 or 700 Processors are used to control a common I/O system. This ensures that the operation of a critical control function is not disrupted by loss of a single processor. Refer to Figure 1. The processors communicate through SY/MAX Class 8030 Type CRM230 LTI modules. Communication to the remote I/O occurs through SY/MAX Class 8030 Type CRM232 RTI modules.

In this type of system, one of the controllers is referred to as the primary and the other as the backup. If the primary processor should fail, system control is transferred to the backup processor which then assumes primary status. When the failed processor (previously the primary) is returned to service, it assumes backup status and the system is again fully operational. Refer to Technical Overview TO-251-XX for additional information on the SY/MAX Transfer Interface System.

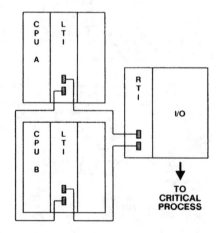

Figure 1
Basic SY/MAX Transfer Interface System

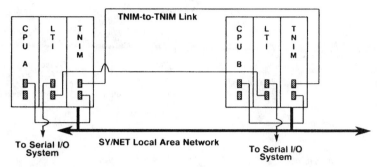

Figure 2
TNIM System Configuration

TNIM System Configuration and Operation

When used in a transfer system, a TNIM resides in the same rack assembly as the processor and LTI. Refer to Figure 2. Each TNIM has a thumbwheel switch which is used to designate a network device number (NDN). The NDN for each TNIM must be selected as an even/odd pair.

The main function of the TNIM pair is to maintain the communication route of the primary controller (primary route) even after system transfer. The primary route is always one greater than that of the odd numbered TNIM. The TNIM pair differs from a SY/NET Class 8030 Type CRM510 Network Interface Module by maintaining the primary communication route.

The TNIM also allows communication instructions to be directed to each module's network device number. This allows data and/or program changes to occur to the backup processor during system operation.

The basic operation of the TNIM pair is as follows:

1) Upon system power-up, both TNIMs in the pair assume backup status.

2) Under command of the current primary controller the associated TNIM is directed to assume primary communication status.

3) Before assuming primary status the directed TNIM:

   a) assumes the transition state to make sure both TNIMs are not in primary status,
   b) checks the status of the paired partner to make sure it is properly configured, (NDN, baud rate, etc.)
   c) commands the partner to assume backup status, and after the new backup TNIM has acknowledged this command.

When a backup TNIM is directed to assume the primary state, the backup TNIM attempts to communicate to the primary TNIM. The backup TNIM first attempts to communicate via the network port. If this is not successful, the backup TNIM uses the TNIM-to-TNIM link.

If the primary processor fails during normal system operation, the same sequence of events occurs as during power-up. In all cases, the transfer between the primary and backup TNIM is initiated under the supervision of the controller which assumes primary status.

TNIM Transfer Implementation

TNIM transfer is implemented by the current primary processor performing a priority WRITE or ALARM communication command to TNIM internal register 20 with a value of one (1). The current state of each TNIM can also be read from register 20. The operational states and their values are:

$$0 = \text{Backup}$$
$$1 = \text{Transition}$$
$$2 = \text{Primary}$$

The TNIM transfer can be initiated in one or the following ways:

1) Upon loss of the primary processor, a flag is set in the backup processor instructing it to assume primary control (SY/MAX LTI/RTI configuration). The new primary processor informs the associated TNIM (current backup) to switch to primary status which in turn set the primary TNIM to backup status.

2) The backup system polls the primary system on a periodic basis for failure (in the primary processor or primary TNIM communication). When a failure is detected, the backup processor assumes primary control and directs the backup TNIM to switch to primary status (non-LTI/RTI configuration).

In a polling configuration, a pre-designated primary and backup device are present. The backup periodically polls the primary to determine if it is still performing its control function (in RUN) and is communicating to other devices on the network. During the polling operation, when the backup device determines the primary system is non-operational (in HALT) or not communicating, the backup would assume to switch states (primary to backup, backup to primary). The new primary controller does this by executing a priority WRITE or ALARM command to backup TNIM register 20 with a value of 1.

SPECIFICATIONS

Class 8030 Type CRM580
Transfer Network Interface Module

Voltage
  Requirement . . . . . 5 VDC, from SY/MAX rack

Current Draw on
SY/MAX Power
  Supply . . . . 950 mA (100% duty cycle)

Operating
  Temperature . . . . . . 32 to 140°F
                          (0 to 60°C)

Storage
  Temperature . . . . . . -40 to +176°F
                          (-40 to +80°C)

Humidity . . . . 5-95% RH, non-condensing

COMM Port . . . . Two independent 9-pin
                          RS-422 ports. COMM "0"
                          must be used for the TNIM-
                          to-TNIM link at 9600 Baud.
                          COMM "1" is standard SY/MAX
                          protocol, and can be set to 300,
                          1200, 2400, and 9600 Baud.

Network Port . . . . . . One dual concentric twist lock
                          connector for the network

Operating Mode . . . . . TRANSFER

Thumbwheel . . . . Sets Network Device Number,
                          valid range is "00" to "99"

Dimensions
  (WxHxD) . . . . 1.5 x 12.8 x 6.6 in.
                          3.8 x 32.5 x 16.8 cm.

Weight . . . . 2.7 lbs. (1.22 kg.)

SY/NET Network

Cable . . . . Baseband twinaxial type, Belden
                          9272 or equivalent

Clock Frequency . . . . 1.0 MHz maximum

Transmission
  Format . . . . Manchester encoded HDLC

Message Format . . . . Packet type, similar to ANSI
                          3.28

Collision Avoidance . . Token-passing based on time
                          windows

Transmission
  Priority . . . . Determined by module's
                          network interface number – the
                          lower the number, the higher
                          that module's priority

Error Checking . . . . . . Cyclic Redundancy Code (CRC)
                          on each transmission

Number of Devices . . . Maximum of 200

Expansion
  Capability . . . . In NET-to-NET mode,
                          multiple networks can be tied
                          together (not applicable to
                          CRM580)

Protection . . . . TransZorbs® Surge Protection on
                          all ports

ORDERING INFORMATION

CLASS	TYPE	DESCRIPTION
8030	CRM580	Transfer Network Interface Module

For additional information refer to Instruction Bulletin 30598-702-01.

Glossary

Definitions of Commonly Used PC Terms

All the new terms introduced in the text are listed here, providing a convenient reference source for those just beginning to learn about programmable controllers.

absorptive law   The following law of Boolean algebra:

$A(A + B) = A$.

ADC   Analog-to-digital converter. *See* analog interface.

address bus   Bus used to gain access to a particular location in memory.

algorithm   *See* user program.

analog interface   An I/O interface that converts an analog signal to digital or vice versa and referred to analog-to-digital or digital-to-analog converter.

AND   A logic operation that requires that all inputs be true for the output to be true.

application memory   The portion of PC memory architecture that contains data memory and user memory.

application program   *See* user program.

architecture   A specification of how various memory systems are organized and used by the PC to perform control functions.

armature   The moving, metallic component of a relay.

ASCII   Acronym for American Standard Code for Information Interchange, a binary code used in computer and PC keyboard applications.

ASCII interface   An interface that allows the transmission of alphanumeric data between the PC and peripherals using the ASCII code.

assembly language   A symbolic programming language not often used for user programs.

associative laws   Two laws of Boolean algebra, as follows:

1. $A + (B + C) = (A + B) + C$
2. $A(BC) = (AB)C$

auxiliary power supply   *See* remote power supply.

base   The number of digits in a number system. Base 10 has ten digits (0 through 9), base 2 has two (0 and 1).

baseband   A type of cable that allows the transmission of only one signal at a time.

BASIC   A high-level, English-statement-type programming language for computers and PCs.

basic language   *See* low-level language.

battery backup system   A battery-based power supply used to protect volatile memories from an interruption in power supply.

baud rate   The number of bits-per-second of binary data that can be received or transmitted during serial communication.

Baudot code   A binary code used primarily in teleprinter machines.

BCD   *See* binary-coded-decimal.

binary   The base-2 number system. It has only two digits: 0 and 1.

binary-coded decimal   A code that represents each digit from 0 to 9 as a four-bit binary number.

binary codes   Systematic methods for presenting or transmitting binary numbers. These include binary-coded decimal (BCD), the Gray code, the Baudot code, and ASCII.

bit   A digit in the binary number system (from *bi*nary digi*t*).

block diagram language   A high-level PC programming language that employs functional block symbols in a ladder format.

Boolean algebra   A mathematical shorthand that describes the outcome of logic operations and combinations of logic operations.

Boolean language   A low-level PC programming language based on symbolic representations of Boolean operations (AND, OR, NOT, NAND, and NOR). It is sometimes called *mnemonic* language.

Boolean operations   Logic operations such as AND, OR, NOT, NAND, NOR, and XOR.

broadband   A type of cable that allows the transmission of two or more signals simultaneously on different channels, or frequencies. The cable used for cable TV is broadband cable.

broadcast   A form of polling in which the master PC sends a request to transmit data to all slave PCs simultaneously and then polls the individual slaves for responses.

buffer   A short-term memory for the temporary storage of data.

bus   A line or group of lines used for data transmission or power distribution.

business-systems local area network   A local area network used to connect office equipment, such as word processors, to a central computer. This type of network allows communications between word processors and computer terminals that are scattered about in various office locations.

byte   A group of eight bits.

C   A high-level programming language.

cable   A shielded, very uniform, wire-based communications medium.

central processing unit   A portion of PC hardware that contains a power supply, micro-processor, and memory. The central processing unit interprets and executes programmed instructions.

chip   Slang for integrated circuit.

clock   A timing device located in the CPU that provides synchronizing pulses.

CMOS   Acronym for complementary metal-oxide semiconductor; a semiconductor technology often used in RAM memory.

CMP   In programming languages, an instruction that requires that two numbers or words be compared. It usually is followed by =, > or <.

coaxial cable   *See* cable.

code   A systematic method for presenting or transmitting information.

coil   The electromagnet portion of a relay. Also, in ladder diagrams, a symbol for an output.

collision detection   A method of controlling network communications popular in business systems networks. Collision detection circuitry senses transmissions on the bus and prevents PCs from transmitting when a signal is on the bus.

common bus   A configuration or topology for serial communication between several PCs.

communication standard   *See* protocol.

communications medium   The physical form that a bus takes (e.g., twisted pair conductors or coaxial cable).

communications network   A communications system linking several PCs. The objective of a communications network is the sharing of data and system status.

commutative laws   Two laws of Boolean algebra, as follows:

1. $A + B = B + A$
2. $AB = BA$

complement   *See* method of complements.

composite logic operations   Logic operations formed from combinations of the three basic logic operations, AND, OR, and NOT. Composite logic operations include NAND, NOR, and XOR.

computer-type language   Any high-level, English-statement-type PC programming language (e.g., BASIC).

contact-ladder diagram   A schematic shorthand notation that represents the current flow status of control relays and associated equipment.

control contact   In ladder diagram programming language, a contact that, when closed, allows the completion of the instructions in the rung.

control network   A communications system linking a PC and peripherals. The objective of a control network is the complete transmission of data and instructions within one scan of the PC.

control program   *See* user program.

control relay   A relay designed for industrial control applications. These are more rugged in design than other relays and almost always are double-throw relays (i.e., they contain both N.O. and N.C. contacts). The armature of a control relay usually contains a pair of moving contacts to help dissipate transient voltages.

core memory   A nonvolatile RAM consisting of small magnetic donuts.

CORONET   A proprietary local area network from Cegelec Industrial Controls.

counter   A programming symbol used to energize or de-energize an output after a specified count of events.

cps   Abbreviation for characters per second. *See* printing speed.

CPU   *See* central processing unit.

CRT   Acronym for cathode ray tube. The CRT is a combination VDT and keyboard that allows system monitoring and program entry. Two main types of CRTs exist: dumb (no microprocessor or memory) and intelligent (containing microprocessor and memory).

DAC   Digital-to-analog converter. *See* analog interface.

daisy chain   A configuration or topology for serial communication between a PC and several peripherals or between several PCs.

data bus   A bus used to transmit data previously stored in a particular location in memory.

data highway   *See* local area network.

Data Highway II   A proprietary local area network by Allen-Bradley Co.

data memory   The portion of PC memory that holds data used by the microprocessor in fulfilling its control functions (e.g., preset values). *Also see* RAM.

data processing interface   An intelligent interface that performs prescribed calculations on data received from sensors or other devices connected to the interface.

debugging   The process of locating and correcting errors in a user program.

decimal   The base-10 number system. It has ten digits, 0 through 9.

DeMorgan's Laws   Two laws of Boolean algebra, as follows:

1. $\overline{(A + B + C)} = \overline{A}\,\overline{B}\,\overline{C}$
2. $\overline{(ABC)} = \overline{A} + \overline{B} + \overline{C}$

diagnostic   *See* diagnostic program.

diagnostic program   A program included in the microprocessor that detects failure in communications, system operation, etc., and activates an alarm circuit to signal a failure.

dibble-dabble   A method of converting numbers from the decimal to the binary number system.

discrete interface   An I/O interface used with devices that provide or require discrete (on or off) signals.

distributed control   *See* multiprocessing.

distributed polling   *See* token passing.

distributed processing   *See* multiprocessing.

distributive laws   Two laws of Boolean algebra, as follows:

1. $A(B + C) = AB + AC$
2. $A + BC = (A + B)(A + C)$

documentation   The process of preparing a record of the user program, hardware connections, etc., for future reference.

drop cable   A relatively short length of cable that connects a peer PC modem to a bus.

duplex transmission   *See* full-duplex transmission.

EAROM   Acronym for electrically alterable read-only memory; a ROM memory that can be erased quickly by application of a relatively low voltage to a pin on the chip.

EEPROM   Acronym for electrically erasable programmable read-only memory; memory erased with an electric charge.

EIA   Acronym for the Electronic Industries Association.

electromagnetic interference   Spurious signals and noise generated by electromagnetic equipment (e.g., pumps, motors, etc.).

EMI   *See* electromagnetic interference.

END MCR   A program instruction that, in combination with MCR, is used to isolate a portion of a ladder diagram.

END ZCL   A programming instruction similar to END MCR and is used with the ZCL instruction.

EPROM   *See* erasable programmable read-only memory.

erasable programmable read-only memory   A nonvolatile memory that can be erased and reprogrammed.

EXCLUSIVE-OR   *See* XOR.

executive   A supervisory program stored permanently in the PC that supervises the operation of the PC. This supervisory function includes control, data processing, and communication with I/O modules and peripherals.

executive memory   The portion of PC memory that contains the executive program (PROM).

false   An off or 0 state.

feedback   A signal sent to the PC that gives the response of the process or system to the PC's instructions.

fiberoptics   A communications medium based on the transmission of visible light instead of audio signals.

flow chart   A schematic diagram of a user program.

frequency-shift keying   A technique used by modems to convert binary voltages into two-frequency audio signals and vice versa.

FSK   *See* frequency-shift keying.

full adder   A binary adding circuit consisting of two half adders and one OR gate used to add three digits.

full-duplex transmission A form of data transmission that allows simultaneous, two-way communication between a PC and peripheral.

gate *See* logic gate.

GEMSMART Intelligent I/O modules from Cegelec Industrial Controls.

GEnet Factory Local Area Network A proprietary local area network from GE Fanuc Automation.

GET A programming instruction that requires the microprocessor to obtain data stored in a particular register.

GOSUB A computer language instruction equivalent to the jump-to-subroutine ladder instruction.

GOTO A computer language instruction equivalent to the jump ladder instruction.

GRAFCET A high-level, functional flow chart programming language.

Gray code A binary code that requires a change in only one bit between successive numbers; used primarily in mechanical-to-electrical conversions (e.g., in positioning applications).

half adder A binary adding circuit consisting of one XOR and one AND gate used to add two digits.

half-duplex transmission A form of data transmission that is two-way (i.e., from the PC to the peripheral and vice versa), but with data transmission occurring only in one direction at a time.

handshaking signal A pair of signs used in communication between peer PCs. The pair consists of a request-to-transmit signal and a clear-to-transmit signal.

hardware The physical, electrical, electronic, and mechanical devices that form a PC system.

hexadecimal The base-16 number system. It has 16 digits: 0 through 9 and A through F.

high-level language A PC programming language based either on block diagrams or computer languages.

IEEE Acronym for the Institute of Electrical and Electronics Engineers.

industrial relay *See* control relay.

input interface module A collection of electronic circuits that provide the signal conditioning and isolation required to connect (or interface) a field input device to the PC's microprocessor.

input/output status memory The portion of PC memory architecture reserved for current input/output statuses (RAM).

intelligent alphanumeric display A programmable device that can display letters, numbers, messages, and warnings.

intelligent interface *See* intelligent I/O module.

intelligent I/O module A programmable I/O module or interface, usually remote, that performs only one control function, such as PID control.

I/O Abbreviation for input/output.

I/O interface *See* input interface module; output interface module.

I/O module *See* input interface module; output interface module.

inverter A NOT gate.

ISO Acronym for the International Standards Organization.

isolation transformer A transformer installed between the ac power line and the PC. It protects the PC from EMI.

JMP *See* jump.

JSB *See* jump-to-subroutine.

jump A programming instruction that, when activated, requires the microprocessor to skip to the designated program rung.

jump-to-subroutine Similar to jump, this program instruction requires the microprocessor to skip to the designated subroutine.

K A measure of memory size; 1K of memory provides the capacity for storage of 1024 words.

L1 Local Area Network A local area network or data highway from Westinghouse Electric Co.

ladder diagram A schematic, shorthand notation that represents the power feed (relay ladder) and current flow (contact ladder) statuses of control relays and associated equipment. Also, a programming language employing ladder diagrams.

ladder language *See* ladder diagram.

latching relay A retentive relay containing latch and unlatch coils. The contacts remain in the energized position until the unlatch coil is energized.

latch out In ladder-diagram language, a symbol for an output that represents the energizing of a latching relay's latch coil.

LCD *See* liquid crystal display.

least-significant digit (bit) In a multidigit (multibit) number, the digit (bit) farthest to the right.

LED *See* light-emitting diode.

light-emitting diode A semiconductor that emits light when forward-biased.

line printer *See* printer.

linear voltage differential transformer *See* LVDT.

liquid-crystal display A device that forms numbers and characters using reflected light from liquid crystals.

LOAD A Boolean programming instruction that symbolizes the initiation of a rung with N.O. control contacts.

LOAD NOT A Boolean programming instruction that symbolizes the initiation of a rung with N.C. control contacts.

local area network A communications network or system that allows the transmission of data and system statuses between PCs, or between PCs and intelligent devices, at high speeds and over long distances; often called a data highway.

logic circuits   Electronic circuits that perform the three basic logic operations of AND, OR, and NOT.

logic gate   A logic circuit.

loop   A configuration or topology for serial communication between several PCs.

low-level language   A PC programming language based either on ladder diagrams or Boolean (mnemonic) logic.

LVDT   A sensor that provides an output voltage proportional to linear displacement.

M   A measure of memory size; 1M of memory provides the capacity for storage of 1,048,576 words ($1M = 1K \times 1K$).

manufacturing automation protocol   *See* MAP.

MAP   Acronym for manufacturing automation protocol, a protocol for data highways or local area networks.

master PC   *See* master-slave local area network.

master-slave local area network   A local area network controlled by a master PC or computer. Communications between PCs in the network (slaves) are routed through the master PC or computer.

MCR   Abbreviation for master control relay, a program instruction that, with END MCR, is used to isolate a portion of a ladder diagram.

memory   A portion of the PC's CPU that stores data and instructions.

memory burner   A device that burns, or impresses, a program onto EPROM memory chips.

memory systems   The various types of memory available for use with PCs. These are generally listed as volatile memory and nonvolatile memory.

memory utilization map   A schematic representation of memory architecture.

method of complements   A method used by computers and PCs to perform binary subtraction.

microprocessor   The integrated circuit and associated support circuitry that performs mathematical, logic, data processing, diagnostic, and control functions. The microprocessor controls all the activities of the PC.

miniprogrammer   A handheld device used to enter programs into the PC's memory or to edit programs.

minuend   A number from which the subtrahend is to be subtracted.

mnemonic language   *See* Boolean language.

Modbus   A proprietary local area network from Modicon, Inc.

modem   Acronym for modulator-demodulator. It serves as an interface between the PC or intelligent device and a communications medium (wire, cable, etc.). It converts two-level binary voltage from the PC to a two-frequency audio signal for transmission and vice versa.

most-significant digit (bit)   In a multidigit (multibit) number, the digit (bit) farthest to the left.

move A block diagram language instruction that is the equivalent of the ladder instructions GET and PUT.

multidrop A peer-to-peer version of the common bus topology.

multiprocessing The use of several microprocessors to perform control tasks, reducing the time required to implement a program.

NAND A composite logic operation formed from an AND operation followed by a NOT operation.

N.C. *See* normally closed.

negative logic A representation in which a true state is represented by zero voltage and a false state is represented by a positive voltage (+5 volts dc).

network adapter module *See* modem.

nibble A group of four bits.

N.O. *See* normally open.

node A device connected to a local area network.

nonvolatile memory A memory that retains its contents even when the power supply is interrupted.

NOR A composite logic operation formed from an OR operation followed by a NOT operation.

normally closed A set of relay contacts that do not pass current when the relay is energized. Also, in ladder diagram programming language, an instruction that indicates that a signal is required to open a contact.

normally open A set of relay contacts that do not pass current when the relay is de-energized. Also, in ladder diagram programming language, an instruction that indicates that a signal is required to close a contact.

NOT A logic operation that provides a true output for a false input, and a false output for a true input. It is often called an *inverter*.

NOVRAM Acronym for nonvolatile random-access memory. It employs conventional semiconductor RAM and EEPROM, both fabricated on a single integrated circuit.

numerical data interface An I/O interface capable of handling data in the form of multiple bits (e.g., BCD inputs and outputs).

octal The base-8 number system. It has eight digits: 0 through 7.

offline programming A type of programming employing intelligent CRTs or programming devices that allows a program to be written and edited without being connected to the PC.

one-shot contacts A program instruction that closes or opens a contact for one program scan only. A typical application is in unlatching a latched relay.

online programming A type of programming, usually employing dumb CRTs, that involves making program changes while connected to the PC.

OR A logic operation that yields a true output when any one input is true.

out A general symbol in ladder diagram programming language that indicates any output (motor, coil, lamp, etc.).

out NOT In ladder diagram programming language, a symbol that indicates that an output is de-energized when the current path through a rung is completed.

output interface module A collection of electronic circuits that provide the signal conditioning and isolation required to connect, or interface, the PC's microprocessor to a field output device.

overflow indicator A warning light or other alarm used to indicate that the capacity of a register has been exceeded.

parallel transmission A form of data transmission between a PC and its peripherals. In parallel transmission, each bit of data is carried by a separate transmission line, thus allowing rapid, simultaneous transmission of data.

PC *See* programmable controller.

peer PC *See* peer-to-peer local area network.

peer-to-peer local area network A local area network in which each PC (peer) controls its own communications and takes a turn at controlling the network.

peripheral An external device connected to the PC, such as a CRT, printer, etc.

personal computer A small, digital electronic device that can store, retrieve, and process data.

photocell A sensor that produces a voltage proportional to the amount of light detected by the sensor.

PID *See* proportional-integral-derivative control.

PID interface An intelligent interface used to provide PID control for various processes that require continuous, closed-loop feedback control.

PLC *See* programmable logic controller.

polling Sometimes called addressing, the process by which a master PC asks a slave PC for data.

positive logic A representation in which a true state is represented by a positive voltage (+5 volts dc) and a false state is represented by zero voltage.

power supply The portion of the CPU that provides power to the microprocessor and memory and often to I/O modules.

PRE In programming, a symbol for preset. It is used to preset timers and counters.

primary battery A nonrechargeable battery.

printer A device that converts electrical signals into printed numbers and characters.

printing speed The number of characters per second a printer can print.

process control The continuous, automatic control of processes or operations.

process measurement The set of techniques used to obtain the values of process variables to be controlled.

processor *See* microprocessor.

program A series of instructions that, when executed, allows the PC to control processes and machines.

program loader A device used to load or reload a program into a PC's memory. Examples include cassette tape recorders and memory burners.

programmable controller A digital electronic device that meets the three following criteria: It has a programmable memory, in which instructions can be stored; the instructions stored in the memory are used to implement various functions, such as logic, sequencing, timing, counting, and arithmetical; and the various functions are used to control machines or processes.

programmable logic controller A primitive, first-generation programmable controller.

programmable read-only memory A nonvolatile memory that cannot be altered once it has been programmed.

PROM *See* programmable read-only memory.

proportional-integral-derivative control A closed-loop feedback control technique, often called PID control. It uses a control signal proportional to the sum of the error signal, the integral of the error signal, and the derivative of the error signal.

protocol A specification of physical details (e.g., number of lines), signal levels, and timing for the standardization of data transmission between a PC and peripherals.

PUT A programming instruction that requires the microprocessor to store an output in a particular register.

query-response A form of polling in which the master PC asks the slave PC to transmit data and then waits a specified time for a response.

radix The base of a number system.

RAM *See* random-access memory.

random-access memory A volatile, semiconductor memory easy to change (i.e., reprogram). It is used for the storage of input data and application programs.

read-only memory A nonvolatile semiconductor memory. Generally, read-only memory is difficult, if not impossible, to reprogram.

register A location in memory for the temporary storage of data, instructions, and information.

regulator An electronic device that maintains power supply output voltage regardless of load.

relay An electromagnetic switch. It contains an electromagnet (commonly called a coil) used to open or close switch contacts.

relay ladder diagram A schematic, shorthand notation that represents the power feed status of control relays and associated equipment.

relay ladder programming *See* ladder diagram.

relay-type instructions In ladder diagrams, these are programming instructions based on relay control systems (e.g., N.C., N.O., latch, unlatch, coil, etc.).

remote I/O modules I/O modules located long distances from the PC location.

remote power supply A power supply for remote I/O modules.

repeater A device that receives a signal and automatically retransmits (or repeats) the signal in amplified form.

RET A programming instruction that requires the microprocessor to return to the main program after completing a subroutine.

ROM *See* read-only memory.

rung One line of program instructions in the ladder diagram programming language.

safety margin The difference between the value of a variable (e.g., temperature) and the value of the variable that is unsafe (e.g., explosive temperature).

scan A process by which the microprocessor accepts inputs, manipulates data, and updates outputs on a periodic basis.

scan rate *See* scan time.

scan time The length of time required to perform a scan. Usually given in milliseconds (msec or ms) per K of memory.

scientific notation A method of conveniently representing either very large or very small numbers as powers of 10.

scratch-pad memory The portion of PC memory used for the temporary storage of data (RAM).

secondary battery A rechargeable battery.

serial interface A special interface module or cable that provides lines and signals for the serial transmission of data between a PC and peripheral.

serial transmission A form of data transmission between a PC and its peripherals. In serial transmission, all data bits are carried one after another (serially) on a single transmission line.

set point The desired operating point (e.g., desired temperature, desired pressure, etc.) of a system or operation.

seven-segment display An output device that displays numbers, comprising either seven-segment light-emitting diodes or seven-segment liquid crystal displays.

simplex transmission A form of data transmission that allows for one-way communication only (e.g., from a PC to a peripheral).

single-pole double-throw A relay that opens one contact and closes another when energized.

single-pole single-throw A relay that either opens or closes a single set of contacts when energized.

skip Either the MCR or ZCL programming instructions.

slave PC *See* master-slave local area network.

SNAP A proprietary programming language from Automatic Timing & Controls.

software Any program or subroutine used by PC hardware.

SPC Control Language A programming language that is an enhanced form of relay ladder logic from Minarik Electric Co.

SPDT   *See* single-pole double-throw.

SPST   *See* single-pole single-throw.

star   A configuration or topology for serial communication between a PC and several peripherals or between several PCs.

State Language   A high-level programming language from Adatek.

Statement List   A text-based programming language from Westinghouse Electric Co.

STEPS™   Proprietary ladder-like programming language from Active Systems Group, Inc.

subroutine   A self-contained program within a larger program. A subroutine can be used several times during the execution of the main program.

subtrahend   A number to be subtracted from a minuend.

SUCONET Field Bus   A proprietary local area network from Klockner-Moeller Corp.

SY/NET   A proprietary local area network from Square D Co.

system memory   The portion of PC memory architecture that contains executive memory and scratch pad memory.

tee tap   A device used to connect a PC communication cable (drop cable) to the bus.

temperature-activated switch   A switch that changes state (from N.O. to N.C. or vice versa) when a certain temperature change is attained.

thermocouple   A device that produces a dc voltage proportional to temperature.

thumbwheel switch   A 10-position switch that transmits a BCD code corresponding to the number displayed on the switch.

timer   A programming symbol used to energize or de-energize an output after a specified time interval.

token-holding time   The length of time during which one peer PC controls a network.

token passing   A method of passing the control of a local area network from one peer to another.

topology   The configuration or physical arrangement of nodes in a local area network.

transmission media   The physical components that allow communication between PCs or between PCs and intelligent devices. Major components of transmission media are the modem (or network adapter module) and the bus, or communications medium.

triaxial cable   *See* cable.

true   An on or 1 state.

truth table   A table that describes the outputs of a particular logic operation for all possible inputs.

TTL   Acronym for transistor-transistor logic.

twisted-pair conductors   A communications medium comprising two single conductors twisted about each other.

unidirectional transmission   *See* simplex transmission.

unlatch out   In ladder diagram language, a symbol for an output that represents the energizing of a latching relay's unlatch coil.

user-friendly   Slang for easy-to-use.

user memory   The portion of PC memory that contains the user program (RAM, EPROM, EAROM).

user program   A program that provides the instructions for a specific control application.

UV-EPROM   An EPROM memory erased upon exposure to ultraviolet light.

VDT   *See* video display terminal.

video display terminal   A monitor (CRT screen) that displays program or system status.

volatile memory   A memory that loses its contents when power supply is interrupted.

word   A group of one or more bytes.

XOR   The EXCLUSIVE-OR composite logic operation, which is the equivalent of one OR, two AND, and two NOT operations. Its truth table is identical to that of the OR operation except it yields a false output when all inputs are true.

ZCL   Abbreviation for zone control last state; program instruction similar to MCR and holds the outputs within the zone in their last states.

zone   A portion of a ladder diagram isolated by the MCR and END MCR instructions, or by the ZCL and END ZCL instructions.

Bibliography

Because programmable controllers are based on digital electronics, a good understanding of the latter will aid in learning about the former. A very good first primer on the subject of digital electronics, which includes material covered in this book (such as number systems, codes, and Boolean functions) plus much more (such as flip-flops, shift registers, counters, displays, and hands-on projects) is the following:

Hawkins, H.M. *Concepts of Digital Electronics.* Blue Ridge Summit: TAB Books, 1983.

In chapter 10, it was noted that before a process can be controlled, the key variable (or variables) must be measured. The use of transducers to convert process variables into measurable voltage has received considerable study. The interested reader is referred to chapter 5 of the following book:

Hallmark, C. *Electronics Measurements Simplified.* Blue Ridge Summit: TAB Books, 1973.

Electronics Measurements Simplified provides for a discussion of transducer measurements and instrumentation systems. Chapter 5 discusses transducers for the measurement of displacement, force, acceleration, pressure, temperature, light, and radiation. A more complete treatment is given in the following:

Kuecken, J. A. *How to Measure Anything with Electronic Instruments.* Blue Ridge Summit: TAB Books, 1981.

This book includes frequency measurement, counting, ADC, and DAC in addition to the topics described in the Hallmark book. For the last word on this subject, see the following:

Carr, J. J. *Digital Interfacing with an Analog World 2nd Ed.* Blue Ridge Summit: TAB Books, 1987.

To capitalize fully on the power and flexibility of programmable controllers, a knowledge of process control theory is required. One of the best introductions to the subject is the following:

Tedeschi, F. P. *How to Design, Build and Use Electronic Control Systems.* Blue Ridge Summit: TAB Books, 1981.

This book provides the basics of both theory and applications. It includes topics not covered in the present work, such as Laplace transforms, Bode plots, and a general discussion on the stability of control systems. It makes no mention of programmable controllers: the control circuits presented are analog circuits. A more advanced book is the following:

Coughanowr, D. R. and Koppel, L. B. *Process Systems Analysis and Control.* New York: McGraw-Hill, 1965.

It provides an in-depth treatment of process control theory. Of course, it makes no mention of programmable controllers, since PCs were not invented when the book was written.

Throughout the present work, PCs have been compared to computers. Readers familiar with computers may want to investigate the following two books:

Weiss, D. *Microprocessors in Industrial Measurement and Control.* Blue Ridge Summit: TAB Books, 1987.

Dalglish, R. L. *An Introduction to Control and Measurement with Microcomputers.* New York: Cambridge University Press, 1987.

These books are very computer-oriented. For example, the latter spends six chapters on the internal operations of computers and on their structure, but only three chapters on interfacing with the outside world. It also assumes some knowledge of BASIC, Pascal, FORTRAN, or other high-level computer languages. Low-level languages such as ladder and Boolean are not mentioned. It was written primarily for computer-literate scientists and engineers.

Of course, there are books on programmable controllers. The following is a standard, near-classic text:

Jones, C. T. and Bryan, L. A. *Programmable Controllers: Concepts and Applications.* Atlanta: International Programmable Controls, Inc., 1983.

This book is excellent but unfortunately is very difficult for the beginner to use because it lacks much basic material. (The present work was written to remedy that deficiency.) The same comment holds true for the following book:

Wilhelm, Jr., R. E. *Programmable Controller Handbook.* Hasbrouck Heights: Hayden Book Co., 1985.

This book is a massive work covering minute details of PC operations and applications. It is too difficult and overwhelming for the beginner but is an excellent reference source for the practicing technician or engineer. Other PC books include the following:

Cox, R. A. *Technician's Guide to Programmable Controllers.* Albany: Delmar, 1984.

Bryan, E. A. and Bryan, L. A. *Programmable Controller Workbook and Study Guide.* Newchurch, K. Ed. Atlanta: International Programmable Controls, Inc., 1987.

Flora, P. C., Ed. *International Programmable Controllers Directory.* Blue Ridge Summit: TAB Books, 1986.

Gilbert, R. *Programmable Controllers: Practices and Concepts.* Willow

Grove, PA: Intertec Pub., 1986.

_____. *Programmable Controllers: Selected Applications Vol I*. Chicago and Atlanta: Industrial Text Co., 1987.

Johannesson, G. *Programmable Control Systems*. Brookfield, VT: Brookfield Pub. Co., 1985.

Kissell, T. E. *Understanding and Using Programmable Controllers*. Englewood Cliffs: Prentice Hall, 1986.

Many of the books just listed, while too difficult for the beginner on his own, work well in classroom situations. Wilhelm's book, for example, was written as a result of his teaching experience in a community college. Several courses are offered on PCs, ranging in length from two to five days and employing some of the texts mentioned above. International Programmable Controls, Inc. offers one based on the Jones and Bryan book. For more information, write to

Industrial Programmable Controls, Inc.
Attn: PC Training
35 Glenlake Parkway-Suite 445
Atlanta, GA 30328

Engineer's Digest, in cooperation with some domestic PC suppliers, has in the past offered a course based on Gilbert's book. Write to

Engineer's Digest
Walker-Davis Publications, Inc.
2500 Office Center
Willow Grove, PA 19090

The Instrument Society of America offers the T2000 series of courses on instrumentation and control fundamentals, including PCs. Write to

Instrument Society of America
P.O. Box 12277
Research Triangle Park, NC 27709

Be prepared to pay dearly for PC courses: they range in price from $375 to more than $1,000. They do, however, offer the advantage of classroom instruction and hands-on experience.

   The PC world changes very rapidly. To stay on top of it, I recommend the three following periodicals:

Control Engineering
Cahners Publishing Co.
Division of Reed Publishing USA
275 Washington St.
Newton, MA 02158

Chilton's I&CS
Chilton Co.
Chilton Way
Radnor, PA 19089

Engineer's Digest
Walker-Davis Publications, Inc.
2500 Office Center
Willow Grove, PA 19090

Index

About the author

George L. Batten, Jr. has over twelve years experience in industrial research. He received the B.S. degree in Chemistry from Wake Forest University and the Ph.D. degree in Chemical Physics from the University of North Carolina. Dr. Batten holds an FCC Commercial License. He is a Fellow of the American Institute of Chemists, and his original research has been published in the *Journal of Organic Chemistry*, the *Journal of Chemical Physics*, the *Journal of Colloid and Interface Science*, and *Tappi Journal*.

Related titles for Programmable Controllers, 0-07-004214-4

Programmable Logic
INTEL CORPORATION
The burgeoning market of programmable logic has brought a plethora of terms and technologies into the industry. With so many choices, which is the right device to select? This handbook has the answers; it contains data sheets, application notes and technical briefs for Intel's entire PLD family.
ISBN 1-55512-180-2   Paper $25.95

Embedded Microprocessors
Intel Corporation
This handbook includes complete product specifications for the 186 microprocessor family, the industry standard for 16-bit embedded microprocessors.
ISBN 1-55512-204-3   Paper $22.95

Embedded Microcontrollers
Intel Corporation
This handbook contains the technical information for Intel's three industry standard microcontrollers.
ISBN 1-55512-203-5   Paper $22.95

How to Order

 Call 1-800-822-8158
24 hours a day,
7 days a week
in U.S. and Canada

 Mail this coupon to:
McGraw-Hill, Inc.
Blue Ridge Summit, PA
17294-0840

 Fax your order to:
717-794-5291

 EMAIL
70007.1531@COMPUSERVE.COM
COMPUSERVE: GO MH

Thank you for your order!

Shipping and Handling Charges

Order Amount	Within U.S.	Outside U.S.
Less than $15	$3.45	$5.25
$15.00 - $24.99	$3.95	$5.95
$25.00 - $49.99	$4.95	$6.95
$50.00 - and up	$5.95	$7.95

EASY ORDER FORM— SATISFACTION GUARANTEED

Ship to:

Name _____

Address _____

City/State/Zip _____

Daytime Telephone No. _____

ITEM NO.	QUANTITY	AMT.

Method of Payment:

☐ Check or money order enclosed (payable to McGraw-Hill)

Shipping & Handling charge from chart below	
Subtotal	
Please add applicable state & local sales tax	
TOTAL	

☐ Cards   ☐ VISA

☐ MasterCard   ☐ DISCOVER

Account No. ⬜⬜⬜⬜⬜⬜⬜⬜⬜⬜⬜⬜⬜⬜

Signature _____ Exp. Date _____
Order invalid without signature

In a hurry? Call 1-800-822-8158 anytime, day or night, or visit your local bookstore.

Code = BC44ZNA